Technical Assistance Guide
for
Federal Construction Contractors

This Technical Assistance Guide is designed to help Government contractors and subcontractors comply with the Federal laws and regulations that prohibit Government contractors from discriminating in employment and require that they undertake affirmative action to ensure equal employment opportunity in their workforces. It is intended only for Government contractors who have construction contracts or subcontracts, including contractors who have federally assisted construction contracts. It is <u>not</u> intended for Government contractors and subcontractors who hold only non-construction (supply and service) contracts.

The contents of this guide have been designed to help construction contractors and subcontractors:

➢ Understand their contractual obligation to comply with the laws administered by OFCCP;

➢ Understand the role of the OFCCP in enforcing Federal equal employment opportunity and affirmative action laws that apply to Federal contractors and subcontractors;

➢ Develop written Section 503 and VEVRAA affirmative action programs where appropriate;

➢ Implement the affirmative action steps that are described in the Standard Federal Equal Employment Specifications published at 41 CFR 60-4.3; and

➢ Prepare for an OFCCP compliance evaluation.

This Technical Assistance Guide does not create new legal requirements or change current legal requirements. Instead, it reflects the views of OFCCP and is intended to serve as a basic resource document on OFCCP-administered laws. The legal requirements related to equal employment opportunity that apply to Federal supply and service contractors are contained in the statutes, executive orders, and regulations cited in the Guide. Every effort has been made to insure that the information contained in the Guide is accurate and up to date.

Table of Contents

Appendices

Overview of OFCCP Mission and Program

The Office of Federal Contract Compliance Programs (OFCCP) administers and enforces three equal employment opportunity laws that apply to Federal government contractors and subcontractors, including construction contractors: Executive Order 11246, as amended, Section 503 of the Rehabilitation Act of 1973, as amended, and the Vietnam Era Veterans' Readjustment Assistance Act of 1974, as amended (VEVRAA). The OFCCP monitors compliance with these laws primarily through compliance evaluations, during which a compliance officer examines the contractor's affirmative action efforts and employment practices. The OFCCP also investigates complaints filed by individuals alleging employment discrimination under any of these laws.

The OFCCP encourages voluntary compliance and provides technical assistance regarding the requirements of the equal employment opportunity laws it enforces. The OFCCP maintains a national office in Washington, DC, six regional offices, and several district offices within each region throughout the United States. See Appendix D for the addresses and phone numbers of key OFCCP offices.

OFCCP Responsibilities

The OFCCP carries out its enforcement responsibilities by:

➤ Offering technical assistance, including providing training workshops and publications (such as this guide) to federal contractors and subcontractors to help them understand regulatory requirements and the compliance evaluation process;

➤ Conducting compliance evaluations and complaint investigations of federal contractors' and subcontractors' personnel policies and practices;

➤ Forming linkage agreements between contractors/subcontractors and the Department of Labor's employment and training programs, outside organizations, and recruitment sources to help employers identify and recruit qualified employees;

➤ Negotiating agreements, including formal Conciliation Agreements with contractors and subcontractors found in violation of regulatory requirements;

➤ Monitoring contractors' and subcontractors' progress in fulfilling the terms of their conciliation agreements through periodic compliance reports;

➤ And when necessary, recommending enforcement actions to the Solicitor of Labor.

Overview of Laws Administered by OFCCP

The OFCCP is responsible for enforcing Federal laws and regulations that prohibit discrimination and require federal contractors and subcontractors to take affirmative action to ensure that all individuals have an equal opportunity for employment without regard to race, color, religion, sex, national origin, disability, or status as a protected veteran. The OFCCP is responsible for administering:

➤ *Executive Order 11246*, as amended, which prohibits discrimination and requires affirmative action to ensure equal employment opportunity without regard to race, color, sex, religion and/or national origin; and the implementing regulations at *41 CFR Parts 60-1 through 60-50*. Generally, all contractors and subcontractors holding non-exempt Federal and federally assisted construction contracts and subcontracts exceeding $10,000 must comply with Executive Order 11246. The regulations implementing the Executive Order establish different affirmative action requirements for construction and non-construction (supply and service) contractors. While all covered Government contractors and subcontractors, both construction and non-construction, are required to take affirmative action, non-construction contractors that meet the 50 employee/$50,000 contract thresholds are required to develop and maintain a written Executive Order 11246 affirmative action program.

➤ *Section 503 of the Rehabilitation Act of 1973, as amended, (Section 503)*, which prohibits discrimination and requires affirmative action in all personnel practices for qualified individuals with disabilities; and the implementing regulations at *41 CFR Part 60-741*. These requirements apply to contractors and subcontractors with a covered Federal contract or subcontract valued in excess of $10,000. In addition, the regulations implementing Section 503 require that covered contractors and subcontractors with a Government contract or subcontract of $50,000 or more and 50 or more employees develop and maintain a written Section 503 affirmative action program.

➤ *The non-discrimination and affirmative action provisions of the Vietnam Era Veterans' Readjustment Assistance Act of 1974, as amended, 38 U.S.C. 4212 (VEVRAA)*, which prohibit discrimination and require affirmative action in all personnel practices regarding covered veterans. As amended, this statute is no longer limited to veterans from the Vietnam Era. VEVRAA now applies to disabled veterans, Armed Forces service medal veterans, recently separated veterans, and other protected veterans who served during a war or in a campaign or expedition for which a campaign badge has been authorized.

    ➤ For a federal contractor or subcontractor with a contract or subcontract of $25,000 or more entered into before December 1, 2003, the implementing regulations are at *41 CFR Part 60-250.* In addition to prohibiting discrimination, these regulations require that covered contractors and subcontractors with a Government contract or subcontract of $50,000 or more and 50 or more employees develop and maintain a written VEVRAA affirmative action program.

> For a federal contractor or subcontractor with a contract or subcontract of $100,000 or more entered into or modified on or after December 1, 2003, the implementing regulations are at *41 CFR Part 60-300.* In addition to prohibiting discrimination, these regulations require that covered contractors and subcontractors with a Government contract or subcontract of $100,000 or more and 50 or more employees develop and maintain a written VEVRAA affirmative action program. Contractors or subcontractors with contracts entered into prior to December 1, 2003 (and not since modified) <u>and</u> contracts entered into on or after December 1, 2003, are subject to both Parts 60-250 and 60-300.[1]

OFCCP shares enforcement responsibilities with other Federal agencies in the administration of the following laws:

> *Immigration Reform and Control Act of 1986 (IRCA),* requires employers to keep certain records (I-9 forms) for the U.S. Citizenship and Immigration Services (USCIS) that verify their employees' eligibility to work in the U.S. (*i.e.,* proof of citizenship or authorization to work). For an electronic copy of the form see: http://www.uscis.gov/files/form/I-9.pdf.

> *Title I of the Americans with Disabilities Act of 1990 (ADA)*, as amended, prohibits employment discrimination by employers with 15 or more employees against qualified individuals on the basis of disability. The Equal Employment Opportunity Commission (EEOC) has primary authority for enforcing the ADA, but OFCCP is authorized to act as EEOC's agent in processing and investigating ADA complaints falling within the overlapping jurisdiction of Section 503 and Title I of the ADA. 41 CFR Part 60-742.

> *Title VII of the Civil Rights Act of 1964*, as amended, prohibits employment discrimination on the basis of race, color, national origin, sex and religion. In many instances, employment discrimination claims against a government contractor can be brought under both Executive Order 11246 and Title VII. While EEOC has primary authority for enforcing Title VII, OFCCP is authorized to act as EEOC's agent in processing, investigating and resolving the Title VII component of complaints filed with OFCCP under Executive Order 11246 that allege discrimination of a systemic or class nature on the basis of race, color, national origin, sex or religion.

---

[1] These threshold and reporting changes result from the passage of the Jobs for Veterans Act (JVA), P.L. 107-288 (Nov. 7, 2002). JVA also broadened the scope and definition of covered veterans.

Commonly Asked Questions

Which construction contractors and subcontractors are subject to OFCCP administered laws?

A ***construction contractor*** or subcontractor is covered under:

➢ Executive Order 11246 if they have:

- ◆ A Federal construction contract or subcontract of over $10,000;

- ◆ A federally assisted construction contract or subcontract of over $10,000;

- ◆ A construction contract or subcontract of over $10,000 with a Federal non-construction contractor or subcontractor, if the construction contract/subcontract is necessary in whole or in part to the performance of the Federal non-construction contract or subcontract; or

- ◆ Multiple Federal construction contracts or subcontracts of less than $10,000 that, when added together total more than $10,000 within any 12 month period or can reasonably be expected to total more than $10,000 during that time.

➢ Section 503 if they have a construction contract/subcontract in excess of $10,000.

➢ VEVRAA if they have a construction contract/subcontract of $100,000 or more (or a contract entered into prior to December 1, 2003 and not since modified of $25,000 or more).

Federally assisted construction contracts and subcontracts are covered under the Executive Order only, and not under either Section 503 or VEVRAA.

Federal and federally assisted construction contractors and subcontractors who are subject to OFCCP requirements have a contractual obligation to comply with the applicable OFCCP-administered laws governing equal employment opportunity and affirmative action. **Additionally, covered contractors and subcontractors must comply with these regulations at**

all work sites. For example, a company with a Federal construction contract in California must not only comply with OFCCP requirements at the California work site where the Federal contract work is being done, but must also comply with the OFCCP requirements at all of the company's work sites throughout the United States.

Are construction contractors' EEO and affirmative action obligations specified in the contract documents?

Yes. A "Notice of Requirement for Affirmative Action To Ensure Equal Employment Opportunity (Executive Order 11246)" is included in the bid solicitations for all Federal and federally assisted construction contracts and subcontracts in excess of $10,000. The Notice, which is published at 41 CFR 60-4.2, informs the contractor/bidder of the affirmative action requirements imposed under Executive Order 11246, including the specified goals for minority and female participation.

The construction contractor's EEO and affirmative action obligations are also specified in the Government contract. Every covered Government construction contract and subcontract must contain the equal opportunity clause found at 41 CFR 60-1.4(a), which specifies the obligations imposed under Executive Order 11246. Covered federally assisted construction contracts and subcontracts must incorporate the equal opportunity clause found at 41 CFR 60-1.4(b). In addition, covered construction contracts and subcontracts must incorporate the equal opportunity clauses found at 41 CFR Parts 60-250.5, 60-300.5 and 60-741.5, which contain the obligations imposed under VEVRAA and Section 503.

The text of these equal opportunity clauses may be expressly included in each contract or subcontract, or incorporated by reference. Importantly, the equal opportunity clauses are deemed to be a part of every covered construction contract and subcontract even if they are not physically incorporated in the contract documents.

In addition to the equal opportunity clauses, Federal and federally assisted construction contracts and subcontracts in excess of $10,000 must include the "Standard Federal Equal Employment Opportunity Construction Contract Specifications," which are found at 41 CFR 60-4.3. The specifications describe the affirmative action obligations and set forth the specific affirmative action steps the construction contractor must implement in order to make a good faith effort to achieve the goals for minority and female participation that were listed in the bid solicitation.

Note: In the following sections of the Technical Assistance Guide, the term "contract" generally refers to both a *contract and subcontract;* the term "subcontract" generally is not used unless it is necessary to the context. Similarly, the terms "contractor" and "construction contractor" include *subcontractors and construction subcontractors,* as well, unless specified otherwise.

Overview of Construction Contractor Responsibilities

Covered Federal construction contractors must comply with Executive Order 11246, Section 503, and VEVRAA, while federally assisted construction contractors must comply with Executive Order 11246 only.

To comply with Executive Order 11246, contractors must demonstrate good faith efforts to meet their affirmative action goals for the employment of minorities and women in the construction industry. In order to take into account the fluid and temporary nature of the construction workforce, OFCCP does not require construction contractors to develop written affirmative action programs. Instead, OFCCP has established utilization goals based on civilian labor force participation rates, and has outlined in the regulations good faith steps for construction contractors to follow.

The goals, by geographic area, are determined by the Deputy Assistant Secretary, OFCCP, and are issued pursuant to 41 CFR 60-4.6. A "Notice of Requirement for Affirmative Action to Ensure Equal Employment Opportunity" is included in the bid specifications for all Federal and federally assisted construction contracts and subcontracts in excess of $10,000. The Notice sets forth the goals for minority and female participation. The goals are expressed as a percentage of the hours worked by the contractor's aggregate workforce in each trade on all construction work performed in the geographic area, regardless of whether the work is Federal, federally assisted or non-federal. Where a contractor performs construction work in a geographic area located outside the geographic area in which it has a covered contract, it must apply the goals established for the geographic area where the work is actually performed. Goals in this second geographic area also are applicable to both federally involved and non-federally involved construction work in that area. See the example in (b) below.

> **(a) Goals for Women:** The current goal for the utilization of women is 6.9% of work hours and applies to all of a contractor's construction sites regardless of where the Federal or federally assisted contract is being performed. This goal was originally published in the Federal Register of April 7, 1978, 43 FR 14899, 14900, as Appendix A. The 6.9% goal was subsequently extended indefinitely, pursuant to a Notice published in the Federal Register of December 30, 1980, 45 FR 85750, 85751.

> **(b) Minority Group Goals:** Goals for minority utilization were first published in the Federal Register of October 3, 1980, 45 FR 65979, 65984, as Appendix B-80. Current goals for the utilization of minorities are listed in Appendix E of this Guide. Minority goals are formulated in terms of work hours performed in a specific Standard Metropolitan Statistical Area (SMSA) or Economic Area (EA). For example, ABC Company has a Federal contract for construction work in SMSA X. The goals for SMSA X apply to all of ABC's construction work in SMSA X, both the federally involved and the non-federally involved construction work. In addition, if ABC Company performs construction work in SMSA Y, it would apply the SMSA Y goals to all its construction work in SMSA Y, whether or not it has a Federal or federally assisted contract in SMSA Y. Although SMSAs were subsequently realigned into "Metropolitan Statistical Areas"

(MSAs) for use in a subsequent census, construction goals continue to be expressed as SMSAs.

These goals are not a requirement for quotas. Quotas are expressly forbidden by law. Affirmative action goals under Executive Order 11246 are targets for recruitment and outreach and should be reasonably attainable by means of applying good faith efforts. The standard of compliance is good faith. Numerical goals do not create guarantees for specific groups, nor are they designed to achieve proportional representation or equal results.

As stated above, construction contractors are **not** required to develop written Executive Order affirmative action programs. In lieu of a written affirmative action program, the regulations enumerate the good faith steps covered construction contractors must take in order to increase the utilization of minorities and women in the skilled trades. These sixteen requirements are discussed in the *Standard Federal Equal Employment Opportunity Construction Contract Specifications* (Executive Order 11246). Construction contractors must document the steps and actions that they take to ensure that these requirements are met. **The specifications are included in covered Federal or federally assisted construction contracts and subcontracts. The specifications are deemed incorporated in all covered contracts by operation of the Executive Order regardless of whether they are incorporated in the solicitation or contract and regardless of whether the contract is written.**

Depending on the size of the construction contractor and the type of relationship it has with the Federal Government, covered construction contractors may have additional responsibilities, such as the following:

a) Including the provisions of the applicable Executive Order 11246, Section 503, and VEVRAA equal employment opportunity clauses in subcontracts and purchase orders;

b) Notifying OFCCP about any construction subcontract awards in excess of $10,000 that are made under covered federal or federally assisted construction contracts;

c) Complying with personnel record retention requirements;

d) Completing and submitting the annual EEO report, Standard Form 100 (also known as the "EEO-1 Report"), if the construction contractor or subcontractor has 50 or more employees and a covered contract or subcontract of $50,000 or more;

e) Complying with the "Uniform Guidelines on Employee Selection Procedures," which are published at 41 CFR Part 60-3;

f) Maintaining a written affirmative action program for qualified individuals with disabilities if the contractor has 50 or more employees and a non-exempt Government contract or subcontract of $50,000 or more;

g) Maintaining a written affirmative action program for covered veterans, if the contractor has 50 or more employees and a non-exempt Government contract or subcontract of $100,000 ($50,000 for a contract covered by the Part 250 regulations);

h)   Completing and submitting the Federal Contractor Veterans' Employment Report using Form VETS-100 or VETS 100A, as appropriate; and

i)   Complying with the Immigration Reform and Control Act (IRCA) of 1986.

Each of the affirmative action program specifications and additional compliance requirements for construction contractors are detailed on the following pages.

Sixteen EEO and Affirmative Action Requirements

The *Standard Federal Equal Employment Opportunity Construction Contract Specifications* (Executive Order 11246), which are published at 41 CFR 60-4.3, require federally-involved construction contractors with a construction contract in excess of $10,000 to take affirmative action steps that are at least as extensive as the 16 affirmative action steps listed in the Specifications. The 16 steps are summarized below. Actions that covered construction contractors are required to take to comply with the steps are included. Examples of suggested or alternative actions that would enable a contractor to comply with the specifications are also listed. The examples listed should not be viewed as being the only possible ways to comply with these specifications. Also, depending on the situation, a contractor may need to take more than one action to comply with the particular specification, as well as take actions that are not specifically listed below.

KEY!
Contractors should document their efforts fully

EEO AND AFFIRMATIVE ACTION SPECIFICATION #1

Contractors and subcontractors must maintain a work environment free of harassment, intimidation, and coercion at all sites and in all facilities at which the contractor's employees are assigned. (41 CFR 60-4.3(a)7.a.)

Contractors must also take specific steps to ensure that all foremen, superintendents, and other on-site supervisory personnel are aware of and carry out the company's contractual obligation to maintain such a working environment, with specific attention to minorities and women working at all work sites and facilities.

Examples of Actions That Demonstrate Compliance:

➤ Contractors may produce and distribute copies of policy statements prohibiting harassment to all employees.

➤ EEO policy statements must be posted at all construction sites.

➤ Contractors may give supervisory personnel and other employees memoranda and other written instructions addressing the need to maintain a work environment free of harassment, intimidation, and coercion. Copies of such written materials should be retained.

➢ Contractors may hold meetings to inform supervisory personnel of their duty to carry out the contractor's obligation to maintain a workplace free of harassment, intimidation, or coercion. Minutes or other records of such meetings should be retained.

➢ Contractors that assign more than one woman to each construction project should retain records of such assignments.

➢ Contractors may develop formal procedures to handle complaints of harassment and maintain records of such complaints and how the company handled them.

➢ Contractors' EEO Officers may prepare and retain reports, diaries, analyses, etc., of specific efforts made to monitor the work environment for the presence of any forms of harassment, intimidation, or coercion, such as: verbal, visual or written abuse; physical aggressiveness; assigning women and/or minorities to more difficult or dangerous work than men/non-minorities; or sabotaging of individual's work.

➢ Contractors may provide harassment awareness training to supervisors or employees. Contractors should retain records of such training that indicate the dates of the training, the names of those conducting the training, the names of those attending the training, and a copy or description of the training materials.

EEO AND AFFIRMATIVE ACTION SPECIFICATION #2

Contractors and subcontractors must establish and maintain current lists of minority and female recruitment sources; provide written notification to minority and female recruitment sources and to community organizations when the contractor or its unions have employment opportunities available; and maintain a record of the organizations' responses. (41 CFR 60-4.3(a)7.b.)

Examples of Actions That Demonstrate Compliance:

➤ Recruitment sources should include the state employment offices serving the recruitment areas for the company's construction projects, and may also include organizations such as the Job Corps, Urban League, YWCA, National Association of Women in Construction, Neighborhood Youth Corps, National Organization of Women, LULAC, and Aspira, among others. In addition, local community organizations are extremely effective as employer/employee linkage resources.

➤ Contractors may maintain files of letters to minority and female recruitment sources announcing the employment opportunities and application procedures. In order to maintain a record of recruitment organizations' responses, contractors may retain any written responses received from the sources, or log or otherwise record the responses.

➤ An applicant flow log may be used by contractors to identify employment solicitations and referrals, and to track the results of the applications. Applicant flow documentation should include copies of correspondence from recruitment sources, copies of job announcements from state employment offices, and copies of notes, diaries, phone logs and/or other written records of contacts with recruitment organizations.

EEO AND AFFIRMATIVE ACTION SPECIFICATION #3

Contractors and subcontractors must maintain current files containing the names, addresses and telephone numbers of each minority or female off-the-street applicant and minority or female referral from a union, recruitment source or community organization and of what action was taken with respect to each individual. Occasionally, contractors/subcontractors will send individuals to the union hiring hall for referral back to the contractor. If the union did not refer the individual to the contractor **or** if the individual was referred but was not hired, the contractor/subcontractor must keep a record of all actions taken, along with the reasons why the referral or hiring did not occur. (41 CFR 60-4.3(a)7.c.)

Examples of Actions That Demonstrate Compliance:

➤ Contractors should establish files that show the names, addresses, telephone numbers and trades of each minority and female applicant and referral.

➤ In addition to an applicant flow log, contractors may wish to note on the actual employment application forms what action was taken with respect to each applicant and the reason for non-hire.

➤ Where an applicant has been referred to the union for referral back to the contractor, contractors should document this action and its results or any follow-up contacts made with the applicant or the union.

EEO AND AFFIRMATIVE ACTION SPECIFICATION #4

Contractors and subcontractors must immediately notify the Deputy Assistant Secretary in writing when the union or unions with which the contractor/subcontractor has a collective bargaining agreement has not referred a woman or minority individual sent by the contractor/subcontractor. Similarly, contractors/subcontractors must notify OFCCP when the contractor/subcontractor has other information that the union referral process has impeded the contractor's efforts to meet its EEO and affirmative action obligations. (41 CFR 60-4.3(a)7.d.)

Examples of Actions That Demonstrate Compliance:

➤ Contractors should keep copies of all letters to and from the unions, minutes of meetings, etc., related to any claims that the union has impeded the company's efforts to comply with its EEO obligations.

➤ Contractors should also keep copies of any letters sent to the OFCCP that contain claims of non-referral or claims that a union has impeded the contractor's efforts to comply with EEO obligations.

Note: *Neither the provisions of a collective bargaining agreement, nor the failure by a union with whom the contractor has a collective bargaining agreement, to refer either minorities or women shall excuse the contractor's obligations under the contract specifications, Executive Order 11246, as amended, or the applicable regulations (see 41 CFR 60-4.3(a)5.).*

EEO AND AFFIRMATIVE ACTION SPECIFICATION #5

Contractors and subcontractors must develop on-the-job training opportunities or participate in training programs for the job area(s) which expressly include minorities and women. Contractors' actions must include upgrading programs, apprenticeships and trainee programs relevant to the contractor's employment needs, especially those programs approved by the Department of Labor. Contractors and subcontractors must provide notice of these training opportunities and job programs to recruitment sources, state employment offices and other referral sources that the contractor/subcontractor has compiled under *Specification 2* above. (41 CFR 60-4.3(a)7.e.)

Examples of Actions That Demonstrate Compliance:

➤ Contractors may maintain records of employees' participation in training programs, including those that are approved or funded by the Department of Labor's Bureau of Apprenticeship and Training.

➤ Contractors may document any contributions of cash, equipment or personnel provided in support of training or apprenticeship programs.

➤ Contractors may inform minority and female recruitment sources and schools of these programs in writing. Contractors should retain copies of any such letters.

EEO AND AFFIRMATIVE ACTION SPECIFICATION #6

Contractors and subcontractors must disseminate EEO policies by:

➤ Providing notice of the policies to unions and training programs and requesting their cooperation and assistance in meeting EEO obligations;

➤ Including EEO policy statements in all policy manuals and collective bargaining agreements;

➤ Publicizing these policies in company newsletters, the annual report, etc.;

➤ Specifically reviewing the policy with all management personnel and with all minority and female employees at least once a year; and,

➤ Posting the EEO Policy on bulletin boards accessible to all employees at each location where construction work is performed. (41 CFR 60-4.3(a)7.f.)

Examples of Actions That Demonstrate Compliance:

➤ In addition to including EEO policies in all policy manuals, contractors may include EEO policies in employee handbooks provided to each employee when they are hired (if such a handbook exists).

➤ Copies of contractors' EEO policies should be posted on bulletin boards that are accessible to all employees at each location where construction work is performed.

➤ Contractors should document discussions that it has with employees about EEO policies. For example, employees may be asked to sign a receipt for an employee handbook that contains EEO policies. Employees can be asked to sign a form at a new employee orientation indicating that the company's EEO policies have been reviewed with them.

➤ Contractors may also keep copies of letters, memoranda and notices to unions and training programs notifying them of the contractor's EEO policies and requirements and requesting their assistance in meeting those obligations.

➤ Contractors can keep a file containing company newsletters and annual reports which contain descriptions of EEO policies.

EEO AND AFFIRMATIVE ACTION SPECIFICATION #7

At least once a year, contractors and subcontractors must review EEO policies and affirmative action obligations (under these specifications) with all employees having any responsibility for hiring, assignment, layoff, termination or other employment decisions. These EEO policies and affirmative action obligations must be specifically reviewed with on-site supervisory personnel such as superintendents, general foremen, etc., prior to starting construction work at any job site. Contractor/subcontractor personnel must maintain records that identify the time and place of these meetings, persons attending, subject matter discussed and disposition of the subject matter. (41 CFR 60-4.3(a)7.g.)

Examples of Actions That Demonstrate Compliance:

➤ Contractors should have written records (memoranda, diaries, minutes of meetings, etc.) that identify the time and place of these meetings, persons attending, subject matter discussed and disposition of the subject matter.

EEO AND AFFIRMATIVE ACTION SPECIFICATION #8

Contractors and subcontractors must disseminate EEO policies externally by including them in any advertising in the news media (including minority and female news media). Contractors and subcontractors must also provide written notification to and discuss EEO policies with other contractors and subcontractors with whom the contractor/subcontractor does or anticipates doing business. (41 CFR 60-4.3(a)7.h.)

Examples of Actions That Demonstrate Compliance:

➤ Contractors should have copies of any employment advertisements or job announcements which specifically include the EEO "tagline." The tagline may state that the contractor is "an equal opportunity employer," or it may alternatively state that all qualified applicants will receive consideration for employment without regard to race, color, religion, gender, or national origin. The tagline should appear in all advertisements placed in media, including those targeted towards minority and female audiences.

➤ Contractors should maintain copies of correspondence with subcontractors that notify them of EEO contractual obligations and the contractor's commitment to compliance.

➤ Contractors should document meetings with construction industry associations and organizations where the Federal EEO and affirmative action contract obligations and methods for facilitating compliance have been discussed or acted upon.

EEO AND AFFIRMATIVE ACTION SPECIFICATION #9

Contractors and subcontractors must direct recruitment efforts, both oral and written, to minority, female and community organizations, to schools with minority and female students, and to minority and female recruitment and training organizations serving the contractor's recruitment area and employment needs. Contractors/subcontractors must send notice to its recruitment sources for women and minorities announcing acceptance of applications for apprenticeship or other training. This notice must be sent no later than one month before publication of apprenticeship and training announcements. Notices must describe the openings, screening procedures and tests to be used in the selection process. (41 CFR 60-4.3(a)7.i.)

Examples of Actions to Demonstrate Compliance:

➢ Contractors should have written records of contacts (such as written communications, telephone calls or personal meetings) with minority and female community organizations, recruitment sources, schools and training organizations. Records should specify the date of the contact, the individual contacted, results of the contact, and any follow-up efforts.

➢ Contractors should also document their contacts with local offices of the state employment service, Private Industry Council, vocational/technical schools or high schools with construction related training programs, Displaced Homemaker Programs, Urban League or Opportunities Industrialization Center (OIC) training and referral programs, or other community based organizations.

➢ If a union is responsible for acceptance into the training programs, contractors should ensure that information is obtained from the union on individuals who were referred from the recruitment sources/organizations that were accepted in the program.

➢ Contractors should maintain records of written contacts to recruitment sources announcing training and apprenticeship opportunities. Recruitment sources must be notified one month before the company begins accepting applications.

EEO AND AFFIRMATIVE ACTION SPECIFICATION #10

Contractors and subcontractors must encourage current minority and female employees to recruit other minority persons and women and, where reasonable, provide after school, summer and vacation employment to minority and female youth both at the work site and in other areas of the contractor's work force. (41 CFR 60-4.3(a)7.j.)

Examples of Actions That Demonstrate Compliance:

➢ Contractors may have copies of diaries, telephone logs or memos indicating contacts (both written and oral) with minority and female employees requesting their assistance in recruiting other minorities and women, and records of the results. Contractors should specifically discuss recommendations for referral with minority and female trade employees.

➢ Supervisors and crew leaders may keep a log of worker referrals from minority or female employees or recruitment sources.

➢ Contractors that provide after-school, summer and vacation employment to minority and female youth should maintain records of such employment. Contractors may also retain on file any letters and other documentation of contact with recruitment sources or local state employment agencies regarding these youth employment programs.

Contractors and subcontractors must validate all tests and other selection requirements where there is an obligation to do so under 41 CFR Part 60-3, the "Uniform Guidelines on Employee Selection Procedures (1978)." (41 CFR 60-4.3(a)7.k.). Actions for demonstrating compliance vary by the number of people employed by the contractor. (41 CFR 60-3.15A(1)).

Examples of Actions That Demonstrate Compliance:

➢   Contractors with 100 or fewer employees who are not required to file an EEO-1 Report should collect data to help determine if the test or selection requirement has a possible adverse impact on any race, sex, or ethnic group (see 41 CFR 60-3.15A(1)). These contractors should maintain and have available records showing, for each year:

   1)   The number of persons hired, promoted and terminated in each trade (*e.g.*, carpenter, brick masons, concrete finishers, ironworkers, mechanics, equipment operators), by sex (gender), and where appropriate, by race and national origin[2];

   2)   The number of applicants for hire and promotion by trade and sex, and where appropriate, by race and national origin; and

   3)   The selection procedures used (such as standardized testing or unstructured interviews and qualifications review) for each trade.

➢   Contractors with more than 100 employees should maintain the records listed above and maintain records for each job that show whether the total selection process for each job has an adverse impact on either gender or on any of the following race and ethnic groups specified in the OFCCP regulations: Blacks, American Indians, Asians, Hispanics, and whites other than Hispanics[3]. For guidance on adverse impact determinations, see Appendix B of this Guide. Contractors should perform adverse impact analyses at least

---

[2] Records should be maintained for each race and national origin constituting more than two percent (2%) of the labor force in the relevant labor area. Contractors do not need to keep track of race or national origin data if one race or national origin group constitutes more than ninety-eight percent (98%) of the labor force in the relevant area.

[3] OFCCP's regulations regarding the race, ethnicity, and job categories to be used by contractors have not changed to reflect the new categories for race, ethnicity, and job categories required for the EEO-1 Report. However, as a matter of enforcement discretion, OFCCP will not cite any contractor for non-compliance with the Executive Order solely because it utilizes the race, ethnicity, or job categories required by the new EEO-1 Report in records required by OFCCP regulations. Further, OFCCP will accept AAPs and supporting records that reflect the race, ethnicity, and job categories outlined in either 41 CFR Part 60-2 or the new EEO-1 Report. For more information, see OFCCP's Directive regarding the use of race and ethnic categories available online at:
http://www.dol.gov/ofccp/regs/compliance/directives/dirindex.htm

---

once a year for each group that comprises at least two percent of the labor force in the relevant area or two percent of the applicable workforce. Where a total selection process does adversely impact a specific race or ethnic group, contractors should maintain and have available records showing which components of the selection process have an adverse impact. Records regarding individual components of the selection process should be collected for at least two years after the adverse impact has been eliminated. Contractors must validate selection procedures that have an adverse impact in accordance with the Uniform Guidelines.

EEO AND AFFIRMATIVE ACTION SPECIFICATION #12

At least once a year, contractors and subcontractors must inventory and evaluate all minority and female personnel for promotional opportunities. Contractors must also encourage these employees to seek or prepare for, through appropriate training, etc., promotional opportunities. (41 CFR 60-4.3(a)7.1.)

Examples of Actions That Demonstrate Compliance:

➢ Contractors may keep written records (memoranda, letters, personnel files, etc.) showing promotional opportunities for women and minorities are reviewed annually.

➢ Contractors may keep written records documenting that the participation of women and minorities in promotional opportunities is encouraged.

EEO AND AFFIRMATIVE ACTION SPECIFICATION #13
Contractors and subcontractors must ensure that seniority practices, job classifications, work assignments and other personnel practices do not have a discriminatory effect, by continually monitoring all personnel and employment related activities to ensure that EEO policies and contractors' obligations under the contract specifications are being carried out. (41 CFR 60-4.3(a)7.m.)

Examples of Actions That Demonstrate Compliance:

➤ Contractors may use data collected under *Specification 11* to determine if seniority practices, job classifications, work assignments or other personnel practices have an adverse impact on women and minorities.

➤ Contractors may wish to audit or examine existing personnel practices periodically or to convene an EEO task force when developing new personnel practices to ensure that EEO obligations are being adequately addressed and incorporated.

➤ Contractors must ensure current policies are reviewed on a regular basis to identify factors that are not equally applied.

The term "facilities" refers to waiting rooms, work areas, eating areas, time clocks, rest rooms, washrooms, locker rooms, and other storage or dressing areas, parking lots, drinking fountains, recreation or transportation, and housing facilities provided for employees.

Examples of Actions That Demonstrate Compliance:

➢ Contractors should offer adequate toilet and changing facilities to all employees to guarantee privacy between the sexes.

➢ Contractors may compile announcements (*e.g.*, flyers, posters, e-mails) of company sponsored events such as training, parties or picnics and documentation reflecting that notification has been disseminated equally to all employees.

EEO AND AFFIRMATIVE ACTION SPECIFICATION #15
Contractors and subcontractors must document and maintain records of all solicitations of offers for subcontracts from minority and female construction contractors and suppliers, including circulation of solicitations to minority and female contractor associations and other business associations. (41 CFR 60-4.3(a)7.o.)

Examples of Actions That Demonstrate Compliance:

➢ Contractors should keep letters or other direct solicitations for subcontracts from minority or female contractors with a record of the specific responses and any follow-up activities done to obtain price quotations.

➢ Contractors may have a list of subcontracts they have awarded to minority or female contractors or suppliers, showing the dollar amounts involved.

➢ Contractors should retain copies of solicitations sent to minority and women's contractor associations or other business associations and state or local governmental agencies.

EEO AND AFFIRMATIVE ACTION SPECIFICATION #16
At least once a year, contractors and subcontractors must conduct a review of all supervisors' adherence to and performance under the company's EEO policies and affirmative action obligations. (41 CFR 60-4.3(a)7.p.)

Examples of Actions That Demonstrate Compliance:

➤ Contractors may keep copies of performance evaluations, memoranda, letters, reports, and minutes of meetings or interviews with supervisors and management personnel about their employment practices as they relate to EEO policy and affirmative action obligations.

➤ Contractors should also compile any written evidence that supervisors and managers have been notified when their employment practices adversely or positively affected the company's EEO and affirmative action posture.

Additional Requirements

ADDITIONAL REQUIREMENTS - CONTRACT CLAUSES
Contractors or subcontractors with Federal construction contracts or subcontracts must include or reference the following clauses in certain (depending on dollar amount) subcontracts and purchase orders resulting from the contract: ➤ Executive Order 11246 equal opportunity clause; ➤ Executive Order 11246 contract specifications clause; ➤ VEVRAA equal opportunity clause; and ➤ Section 503 equal opportunity clause. Federally assisted construction contracts or subcontracts must include or reference the following clauses in certain (depending on dollar amount) subcontracts and purchase orders resulting from the contract: ➤ Executive Order 11246 equal opportunity clause; and ➤ Executive Order 11246 contract specifications clause.

Explanation of Requirements:

➤ Federal construction contractors must include or reference provisions of the Executive Order 11246 equal opportunity clause shown in 41 CFR 60-1.4(a) in each subcontract or purchase order of more than $10,000 per year resulting from the contract. Federally assisted construction contractors must include or reference provisions of the Executive Order 11246 equal opportunity clause shown in 41 CFR 60-1.4(b) in each subcontract or purchase order of more than $10,000 per year resulting from the contract.

➤ Whenever a contractor or subcontractor subcontracts a portion of the work involving any construction trade, the Specifications, including the sixteen EEO and affirmative action program requirements described above and the "Notice of Requirement for Affirmative Action to Ensure Equal Employment Opportunity" (41 CFR 60-4.2(d)) containing the applicable goals for minority and female participation must be included in subcontracts larger than $10,000.

➤ Regulations implementing the Vietnam Era Veterans' Readjustment Assistance Act of 1974, as amended (38 U.S.C. 4212), require that contractors include or reference the provisions of the equal opportunity clause for covered veterans in each subcontract and purchase order. See 41 CFR Parts 60-250.5 and 300.5.

Note: Contractors holding federally assisted contracts are not covered by VEVRAA, and are not required to include this clause.

➤ Regulations implementing Section 503 of the Rehabilitation Act of 1973, as amended, require that contractors include or reference the equal opportunity clause for individuals with disabilities (at 41 CFR 60-741.5) in each subcontract and purchase order in excess of $10,000.

➤ Note: Contractors holding federally assisted contracts are not covered by Section 503, and are not required to include this clause.

ADDITIONAL REQUIREMENTS - NOTIFICATION OF AWARD
Contractors and subcontractors must notify OFCCP in writing within 10 working days of the award of any construction subcontract in excess of $10,000 that is made under a covered Federal or federally assisted construction contract. Contractors/subcontractors may fulfill this requirement by notifying the nearest OFCCP district office. (See Appendix D of this Guide.)

Explanation of Requirements:

➤ Per 60-4.2d(3), written notification must include:

- Name, address and telephone number of the subcontractor;

- The subcontractor's employer identification number;

- Estimated dollar amount of subcontract;

- Estimated starting and completion dates of the subcontract; and

- Geographic area in which the subcontract is to be performed.

ADDITIONAL REQUIREMENTS - RECORDKEEPING

Contractors and subcontractors must keep records about their entire on-site construction trade workforce within each covered area in which they perform any construction work (both Federal and non-Federal).
See 41 CFR 60-1.12(a) and 60-4.3(a)14.

Explanation of Requirements:

➤ A covered area (also referred to as a geographical area) is the area identified in the solicitation that generated the Federal or federally assisted construction contract or subcontract.

➤ Any personnel or employment record made or kept by the contractor must be preserved. Federal construction contractors with 150 or more employees and a Government contract of at least $150,000 must preserve such records for no less than two years from the date of making the record or the date the personnel action occurred, whichever is later. Federally assisted construction contractors, and federal construction contractors with fewer than 150 employees or a Government contract of less than $150,000, must retain such records for a minimum of one year from the date they were created or the date the personnel action occurred, whichever is later.

➤ Relevant records include, but are not necessarily limited to, records pertaining to hiring, assignment, promotion, demotion, transfer, layoffs, terminations, rates of pay or other terms of compensation, selection for training and apprenticeship, results of physical examinations (kept in a confidential medical file), job postings, job advertisements, applications, resumes, tests, test results, and job interview notes.

➤ Contractors must keep records that include, at a minimum for each employee, the name, address, telephone number, social security number, race, gender, rate of pay, construction trade, job title (for example, "Equipment Operator," "Apprentice Trainee," "Laborer"), dates of change in job status, hours worked per week in each indicated trade, locations at which the work was performed, union affiliation if any, and employee identification number if any. The records must be maintained in an easily understandable and retrievable form. However to the extent that existing records satisfy this requirement, contractors are not required to maintain separate records.

➤ Contractors may transfer their original paper records to an electronic recordkeeping system, if the medium used accurately reproduces the paper original and would constitute a duplicate or substitute copy of the original paper record under Federal law.

ADDITIONAL REQUIREMENTS - EEO-1 REPORT

Contractors and subcontractors with 50 or more employees and with a covered contract or subcontract of $50,000 or more must submit an annual EEO-1 Report (41 CFR 60-1.7a).

Explanation of Requirements:

➤ The EEO-1 Report (which identifies employees in job categories by race, ethnicity, and sex) is sent to the Joint Reporting Committee (JRC), which is comprised of representatives from the Department of Labor and the Equal Employment Opportunity Commission. Reports must be filed with the JRC by September 30.

➤ Contractors that maintain a single establishment must only complete one EEO-1 Report yearly.

➤ Contractors that maintain multiple establishments must file:

  ♦ One report covering the company's principal or headquarters office;

  ♦ A separate report for each establishment employing 50 or more people;

  ♦ A consolidated report for the entire company which includes all employees.

➤ The EEO-1 Joint Reporting Committee can be reached at: P.O. Box 19100, Washington, DC 20031-9100; by calling 1-866-286-6440; or by e-mailing e1.techassistance@eeoc.gov.

A sample EEO-1 Report form is provided on the next two pages.

Joint Reporting
Committee
- Equal Employment
Opportunity Com-
mission
- Office of Federal
Contract Compli-
ance Programs (Labor)

EQUAL EMPLOYMENT OPPORTUNITY

EMPLOYER INFORMATION REPORT EEO—1

Standard Form 100
REV. 01/2006

O.M.B. No. 3048-0007
EXPIRES 01/2009
100-214

Section A—TYPE OF REPORT
Refer to instructions for number and types of reports to be filed.

1. Indicate by marking in the appropriate box the type of reporting unit for which this copy of the form is submitted (MARK ONLY ONE BOX).

(1) ☐ Single-establishment Employer Report

Multi-establishment Employer:
(2) ☐ Consolidated Report (Required)
(3) ☐ Headquarters Unit Report (Required)
(4) ☐ Individual Establishment Report (submit one for each establishment with 50 or more employees)
(5) ☐ Special Report

2. Total number of reports being filed by this Company (Answer on Consolidated Report only)_____

Section B—COMPANY IDENTIFICATION (To be answered by all employers)

OFFICE USE ONLY

1. Parent Company

a. Name of parent company (owns or controls establishment in item 2) omit if same as label

a.

Address (Number and street)

b.

City or town	State	ZIP code

c.

2. Establishment for which this report is filed. (Omit if same as label)

a. Name of establishment

d.

Address (Number and street)	City or Town	County	State	ZIP code

e.

b. Employer identification No. (IRS 9-DIGIT TAX NUMBER)

f.

c. Was an EEO–1 report filed for this establishment last year? ☐ Yes ☐ No

Section C—EMPLOYERS WHO ARE REQUIRED TO FILE (To be answered by all employers)

☐ Yes ☐ No	1.	Does the entire company have at least 100 employees in the payroll period for which you are reporting?
☐ Yes ☐ No	2.	Is your company affiliated through common ownership and/or centralized management with other entities in an enterprise with a total employment of 100 or more?
☐ Yes ☐ No	3.	Does the company or any of its establishments (a) have 50 or more employees AND (b) is not exempt as provided by 41 CFR 60–1.5, AND either (1) is a prime government contractor or first-tier subcontractor, and has a contract, subcontract, or purchase order amounting to $50,000 or more, or (2) serves as a depository of Government funds in any amount or is a financial institution which is an issuing and paying agent for U.S. Savings Bonds and Savings Notes?

If the response to question C–3 is yes, please enter your Dun and Bradstreet identification number (if you have one): ☐☐☐☐☐☐☐☐☐

NOTE: If the answer is yes to questions 1, 2, or 3, complete the entire form, otherwise skip to Section G.

Section D-EMPLOYMENT DATA

Employment at this establishment – Report all permanent full- and part-time employees including apprentices and on-the-job trainees unless specifically excluded as set forth in the instructions. Enter the appropriate figures on all lines and in all columns. Blank spaces will be considered as zeros.

Job Categories	Number of Employees (Report employees in only one category)														
	Race/Ethnicity														Total Col A - N
	Hispanic or Latino		Not-Hispanic or Latino												
			Male						Female						
	Male	Female	White	Black or African American	Native Hawaiian or Other Pacific Islander	Asian	American Indian or Alaska Native	Two or more races	White	Black or African American	Native Hawaiian or Other Pacific Islander	Asian	American Indian or Alaska Native	Two or more races	
	A	B	C	D	E	F	G	H	I	J	K	L	M	N	O
Executive/Senior Level Officials and Managers 1.1															
First/Mid-Level Officials and Managers 1.2															
Professionals 2															
Technicians 3															
Sales Workers 4															
Administrative Support Workers 5															
Craft Workers 6															
Operatives 7															
Laborers and Helpers 8															
Service Workers 9															
TOTAL 10															
PREVIOUS YEAR TOTAL 11															

1. Date(s) of payroll period used: _____ (Omit on the Consolidated Report.)

Section E - ESTABLISHMENT INFORMATION (Omit on the Consolidated Report.)

1. What is the major activity of this establishment? (Be specific, i.e., manufacturing steel castings, retail grocer, wholesale plumbing supplies, title insurance, etc. Include the specific type of product or type of service provided, as well as the principal business or industrial activity.)

Section F - REMARKS

Use this item to give any identification data appearing on the last EEO-1 report which differs from that given above, explain major changes in composition of reporting units and other pertinent information.

Section G - CERTIFICATION

Check one
1. ☐ All reports are accurate and were prepared in accordance with the instructions. (Check on Consolidated Report only.)
2. ☐ This report is accurate and was prepared in accordance with the instructions.

Name of Certifying Official	Title	Signature	Date
Name of person to contact regarding this report	Title	Address (Number and Street)	
City and State	Zip Code	Telephone No. (including Area Code and Extension)	Email Address

All reports and information obtained from individual reports will be kept confidential as required by Section 709(e) of Title VII.
WILLFULLY FALSE STATEMENTS ON THIS REPORT ARE PUNISHABLE BY LAW, U.S. CODE, TITLE 18, SECTION 1001

Covered federal construction contractors and subcontractors must comply with the nondiscrimination and affirmative action requirements of Section 503 and VEVRAA. Additionally, contractors and subcontractors holding a Federal contract of at least $50,000 and having 50 or more employees must prepare a written affirmative action program for qualified individuals with disabilities. Contractors and subcontractors holding a Federal contract of at least $100,000 and having 50 or more employees must prepare a written affirmative action program (AAP) for covered veterans. These written affirmative action programs may be developed separately or combined.

*NOTE: Section 503 and VEVRAA AAP requirements **do not** apply to federally assisted construction contracts.*

Explanation of Requirement:

➢ *Equal opportunity clauses.* Federal contractors must include or reference the Section 503 equal opportunity clause in all subcontracts and purchase orders in excess of $10,000. Federal contractors must include or reference the VEVRAA equal opportunity clause in all subcontracts and purchase orders of $100,000 or more.

➢ *Invitation to self-identify: Section 503.* The contractor shall, after making an offer of employment to a job applicant and before the applicant begins his or her employment duties, invite the applicant to inform the contractor whether the applicant believes that he or she may be protected by Section 503 and wishes to benefit under the contractor's Section 503 affirmative action program. The contractor may also invite self-identification prior to making a job offer only when:

- ◆ The invitation is made when the contractor actually is undertaking affirmative action for individuals with disabilities at the pre-offer stage; or

- ◆ The invitation is made pursuant to a Federal, state or local law, such as Section 503, that requires affirmative action for individuals with disabilities.

For further explanation of self-identification requirements, see 41 CFR Part 60-741.42. There is a sample invitation to self-identify in Appendix B to 41 CFR Part 60-741.

➢ *Invitations to self-identify: VEVRAA .* The invitation to self-identify requirement for disabled veterans mirrors the Section 503 requirement for individuals with disabilities. The contractor may invite disabled veterans to self-identify <u>prior</u> to making a job offer only when:

- The invitation is made when the contractor actually is undertaking affirmative action for disabled veterans at the pre-offer stage; or

- The invitation is made pursuant to a Federal, state or local law, such as VEVRAA, that requires affirmative action for disabled veterans at the pre-offer stage.

The contractor must also invite applicants to inform the contractor whether the applicant believes that he or she is a recently separated or other protected veteran under VEVRAA and wishes to benefit under the affirmative action program. This general invitation to veterans may be extended at any time before the applicant begins his or employment duties.

For further explanation of self-identification requirements, see 41 CFR Parts 60-250.42 and 60-300.42. There is a sample invitation to self-identify at 41 CFR Part 60-300, Appendix B.

➢ *Personnel Practices.* Contractors must review personnel practices to ensure that the qualifications of known protected veterans or individuals with disabilities are given proper consideration for job vacancies filled either by hiring or promotion, and for all training opportunities offered or available.

- If contractors find that any of these practices have been discriminatory, the practice must be changed and the change must be noted in the contractor's affirmative action program.

- Individual personnel actions (including pre-employment testing) should also be carefully documented. Contractors should be able to provide records of every opening for which an individual with a disability or protected veteran had been considered. Personnel records or employment application forms should identify a specific job opening. If a worker or an applicant who is an individual with a disability or a protected veteran was not selected, contractors should provide a comparison of the qualifications of the person selected with those of the individual with a disability or protected veteran. Records should also indicate what accommodations (if any) were considered to enable the disabled or veteran worker to perform the job.

- With respect to protected veterans, contractors may only use those portions of a person's military record that are job-related.

➢ *Mental and Physical Job Requirements.* Contractors must review all mental and physical job requirements used in selection processes and in medical standards, information and qualifications.

- Examples of mental and physical job requirements include job descriptions containing phrases such as "must be able to lift 50 pounds," or "carry heavy mail bags to and from the accounting department," or "must be able to tolerate heights." Other examples may include policy statements about desired weight, height, physical condition, vision, etc. of the employee.

- Except in the following circumstances, it is unlawful under Section 503 and the Americans with Disabilities Act (ADA) of 1990, as amended, 42 U.S.C. 12112(c), for contractors to require a medical examination of an applicant or employee, or to make inquiries as to whether an applicant or employee is an individual with a disability or as to the nature or severity of such disability.

  - The contractor may make pre-employment inquiries into the abilities of an applicant to perform job-related tasks, or may ask the applicant to describe or demonstrate how, with or without reasonable accommodation, they will be able to perform job-related functions.

  - The contractor may require a medical examination (and/or inquiry) after making an offer of employment to a job applicant and before the applicant begins his or her employment duties, and may condition an offer of employment on the results of such examination (and/or inquiry), if all entering employees in the same job category are subjected to such an examination (and/or inquiry) regardless of disability.

  - Post-employment, the contractor may only make a disability-related inquiry of an employee, or require that an employee have or submit to a medical test or examination, if the inquiry, test, or examination is job related and consistent with business necessity.

➢ *Reasonable Accommodation.* Contractors must provide reasonable accommodation to the known physical and/or mental limitations of applicants and employees with disabilities or disabled veterans, unless the contractor can demonstrate that the needed accommodation would impose an undue hardship on the operation of its business.

The term *reasonable accommodation* means:

  (i) Modifications or adjustments to a job application process that enable a qualified applicant who is an individual with a disability or a disabled veteran to be considered for the position such applicant desires; or

  (ii) Modifications or adjustments to the work environment, or to the manner or circumstances under which the position held or desired is customarily performed, that enable a qualified individual with a disability or disabled veteran to perform the essential functions of that position; or

(iii) Modifications or adjustments that enable the contractor's employee who is an individual with a disability or a disabled veteran to enjoy equal benefits and privileges of employment as are enjoyed by the contractor's other similarly situated employees who are not individuals with disabilities or disabled veterans.

- *Reasonable accommodation* may include but is not limited to:

    (i) Making existing facilities used by employees readily accessible to and usable by individuals with disabilities and disabled veterans; and

    (ii) Job restructuring; part-time or modified work schedules; reassignment to a vacant position; acquisition or modifications of equipment or devices; appropriate adjustment or modifications of examinations, training materials, or policies; the provision of qualified readers or interpreters; and other similar accommodations for individuals with disabilities or disabled veterans.

➤ *Employment Practices Review.* Contractors must undertake appropriate outreach and recruitment activities that are reasonably designed to effectively recruit qualified individuals with disabilities, qualified disabled veterans, and other protected veterans. The extent to which a contractor needs to adopt outreach and recruitment efforts depends on all the circumstances, including the contractor's size and resources, and the extent to which existing employment practices are adequate. To comply with this requirement, a contractor may undertake practices such as the following:

- Contractors should develop a system of internal company communications that fosters acceptance and support of the affirmative action program within their company.

- Contractors should develop a system of checks and audits to ensure that affirmative action measures are being fully implemented.

- Contractors should actively recruit applicants who are individuals with disabilities and protected veterans through schools and training institutions, consumer groups, veterans' employment representatives at state employment services, vocational training programs and any other sources that can provide support and assistance (*e.g.*, a state vocational rehabilitation agency).

- Contractors should include individuals with disabilities in consumer, promotional or recruitment advertising.

- Contractors should secure the cooperation and understanding of subcontractors and unions, vendors and suppliers.

- ♦ Contractors should review employees' records to see if their abilities are being fully used.

- ♦ Contractors should review employees' records to determine who is eligible for promotion or transfer.

Note: *This requirement is similar to* <u>*Reviewing Personnel Practices*</u>*, but goes beyond demanding equal opportunity by requesting specific affirmative actions from contractors. This element also applies to a broader range of contractor activities.*

➢ *Mandatory Job Listing.* Contractors are required, under VEVRAA, to list with the State workforce agency job bank or the local employment service delivery system all employment openings except executive and top management positions, those positions that will be filled from within the contractor's organization, and positions lasting three days or less. All employment openings include full-time employment, temporary employment of more than three days' duration, and part-time employment.

ADDITIONAL REQUIREMENTS - VETS 100 AND VETS 100A REPORTS

Once a year, federal contractors and subcontractors covered under VEVRAA must compile a report of the numbers of disabled and other covered veterans in their work force by job category and hiring location. Contractors/subcontractors must also collect data indicating the total number of employees and the number of disabled veterans, Armed Forces service medal veterans, recently separated veterans, and other covered veterans hired during the reporting period. Contractors and subcontractors must use the VETS-100 or VETS-100A form, as appropriate, for this report. *[41 CFR Part 61-250 and Part 61-300.]*

Explanation of Requirements:

➢ A VETS-100 Report is to be completed by each federal contractor or subcontractor with a contract or subcontract of $25,000 or more entered into before December 1, 2003, and not modified after that date.

➢ The VETS-100A Report is to be completed by each federal contractor or subcontractor with a contract or subcontract entered into or modified on or after December 1, 2003, in the amount of $100,000 or more.

➢ Contractors or subcontractors with multiple work establishments must prepare a VETS-100/VETS-100A report for:

  ◆ The company's principal or headquarters office;

  ◆ Each hiring location employing 50 or more persons; and

  ◆ Each hiring location with less than 50 employees or consolidated reports for all hiring locations in each state.

For more information or to request VETS-100 or VETS-100A Report forms, visit the VETS-100/100A website at http://vets.dol.gov/vets100/, e-mail the VETS-100 staff at HELPDESK@VETS100.com or call (301) 306-6752.

A sample form currently in use as of the publication date of this Guide is on the following page. Downloadable forms are available at: https://vets100.vets.dol.gov.

FEDERAL CONTRACTOR VETERANS' EMPLOYMENT REPORT VETS-100A

(For covered contracts entered into or modified on or after December 1, 2003.)

OMB NO: 1293-0005
Expires:

Persons are not required to respond to this collection of information unless it displays a valid OMB number

RETURN COMPLETED REPORT TO:
U.S. DEPARTMENT OF LABOR
VETERANS' EMPLOYMENT AND TRAINING SERVICE
VETS-100 Reporting Office
4200 Forbes Blvd., Suite 202
Lanham, Maryland 20706

ATTN: Human Resource/EEO Department

TYPE OF REPORTING ORGANIZATION (Check one or both, as applicable)	TYPE OF FORM (Check only one)
☐ Prime Contractor	☐ Single Establishment
☐ Subcontractor	☐ Multiple Establishment-Headquarters
	☐ Multiple Establishment-Hiring Local
	☐ Multiple Establishment-State Consc
	(specify number of locations) ____

COMPANY IDENTIFICATION INFORMATION (Omit items preprinted above-ADD Company Contact Information Below)

COMPANY No:	TWELVE MONTH PERIOD ENDING			2	
		M M	D D	Y	
NAME OF PARENT COMPANY:	ADDRESS (NUMBER AND STREET):				
CITY:	COUNTY:	STATE:	ZIP COD		
NAME OF COMPANY CONTACT:	TELEPHONE FOR CONTACT:	EMAIL:			

NAME OF HIRING LOCATION:	ADDRESS (NUMBER AND STREET):		
CITY:	COUNTY:	STATE:	ZIP COD

NAICS:		DUNS:		-		-		EMPLOYER ID (IRS TAX No.)		-	

INFORMATION ON EMPLOYEES

REPORT ALL PERMANENT FULL-TIME OR PART-TIME EMPLOYEES AND NEW HIRES WHO ARE VETERANS, AS DEFINED ON REVERSE. DATA ON NUMBER OF EMPLOYEES IS TO BE ENTERED IN COLUMNS L, M, N, O, AND P, LINES 1-10. DATA ON NEW HIRES IS TO BE ENTERED IN COLUMNS Q, R, S, T, AND U. ENTER THE MAXIMUM AND MINIMUM NUMBER OF EMPLOYEES. INSTRUCTIONS ARE FOUND ON THE REVERSE OF THIS FORM.

JOB CATEGORIES	NUMBER OF EMPLOYEES					NEW HIRES (PREVIOUS 12 MONTHS)				
	DISABLED VETERANS (L)	OTHER PROTECTED VETERANS (M)	ARMED FORCES SERVICE MEDAL VETERANS (N)	RECENTLY SEPARATED VETERANS (O)	TOTAL EMPLOYEES, BOTH VETERANS AND NON-VETERANS (P)	DISABLED VETERANS (Q)	OTHER PROTECTED VETERANS (R)	ARMED FORCES SERVICE MEDAL VETERANS (S)	RECENTLY SEPARATED VETERANS (T)	TOTAL BOTH VE NON-U
EXECUTIVE/SENIOR LEVEL OFFICIALS AND MANAGERS 1										
FIRST/MID LEVEL OFFICIALS AND MANAGERS 2										
PROFESSIONALS 3										
TECHNICIANS 4										
SALES WORKERS 5										
ADMINISTRATIVE SUPPORT WORKERS 6										
CRAFT WORKERS 7										
OPERATIVES 8										
LABORERS/HELPERS 9										
SERVICE WORKERS 10										
TOTAL 11										

Report the total maximum and minimum number of permanent employees during the period covered by this report.

Maximum Number	Minimum Number

VETS-100 FEDERAL CONTRACTOR REPORT ON VETERANS' EMPLOYMENT

OMB NO:1293-0005
Expires: 04/30/2011

Persons are not required to respond to this collection of information unless it displays a valid OMB number

ATTN: Human Resources/EEO Department

RETURN COMPLETED REPORT TO:
U.S. DEPARTMENT OF LABOR
VETERANS' EMPLOYMENT AND TRAINING SERVICE
VETS-100 Reporting Office
P.O. Box 728
Lanham, Maryland 20703-0728

TYPE OF REPORTING ORGANIZATION (Check one or both, as applicable)	TYPE OF FORM (Check only one)
☐ Prime Contractor ☐ Subcontractor	☐ Single Establishment ☐ Multiple Establishment-Headquarters ☐ Multiple Establishment-Hiring Location ☐ Multiple Establishment-State Consolidated (specify number of locations)_____(MSC)

COMPANY IDENTIFICATION INFORMATION (Omit items preprinted above-ADD Company Contact Information Below)

COMPANY No:	TWELVE MONTH PERIOD ENDING						2 0 0 8
		M M D D Y Y Y Y
NAME OF PARENT COMPANY:	ADDRESS (NUMBER AND STREET):
CITY:	COUNTY:	STATE:	ZIP CODE:
NAME OF COMPANY CONTACT:	TELEPHONE AND EMAIL FOR CONTACT:
NAME OF HIRING LOCATION:	ADDRESS (NUMBER AND STREET):
CITY:	COUNTY:	STATE:	ZIP CODE:

NAICS [ ][ ][ ][ ][ ][ ]   DUNS [ ][ ][ ][ ]-[ ][ ][ ][ ]-[ ][ ][ ][ ]   EMPLOYER ID (IRS TAX no.) [ ][ ]-[ ][ ][ ][ ][ ][ ][ ]

EMPLOYEE DATA AND VETERAN REPORTING REQUIREMENTS

REPORT ALL PERMANENT FULL-TIME OR PART-TIME EMPLOYEES AND NEW HIRES WHO ARE TARGETED VETERANS. DATA ON NUMBER OF EMPLOYEES ARE TO BE ENTERED IN COLUMNS L, M, AND N, LINES 1-9. DATA ON NEW HIRES ARE TO BE ENTERED IN COLUMNS O, P, Q, R, AND S. INSTRUCTIONS ARE FURTHER DETAILED ON THE REVERSE OF THIS FORM.

JOB CATEGORIES	NUMBER OF EMPLOYEES			NEW HIRES (PREVIOUS 12 MONTHS)				
	SPECIAL DISABLED VETERANS (L)	VIETNAM ERA VETERANS (M)	OTHER PROTECTED VETERANS (N)	SPECIAL DISABLED VETERANS (O)	VIETNAM ERA VETERANS (P)	NEWLY SEPARATED VETERANS (Q)	OTHER PROTECTED VETERANS (R)	TOTAL NEW HIRE BOTH VETERANS AND NON-VETERANS (S)
OFFICIALS AND MANAGERS 1								
PROFESSIONALS 2								
TECHNICIANS 3								
SALES WORKERS 4								
OFFICE AND CLERICAL 5								
CRAFT WORKERS (SKILLED) 6								
OPERATIVE (SEMI-SKILLED) 7								
LABORERS (UNSKILLED) 8								
SERVICE WORKERS 9								
TOTAL 10								

Report the total maximum and minimum number of permanent employees during the period covered by this report.

Maximum Number	Minimum Number

ADDITIONAL REQUIREMENTS - I-9 FORMS
Under the Immigration Reform and Control Act of 1986 (IRCA), contractors and subcontractors must maintain I-9 forms to verify that their employees are legally authorized to work in the United States.

Explanation of Requirements:

➤ OFCCP will review contractors' records to verify the following actions have been performed to comply with this law:[4]

- ◆ New employees must complete an I-9 form when they start work;

- ◆ Contractors must check documents that indicate the employee's identity (e.g., driver's license, passport) and eligibility to work (e.g., work visa, social security card);

- ◆ Contractors must properly complete the verification sections on the I-9 form;

- ◆ Contractors must keep I-9 forms for at least three (3) years or at least one year after a person leaves the contractor's employment for employees who stay for more than three years; and

- ◆ I-9 forms must be presented to U.S. Citizenship and Immigration Services (USCIS) or DOL investigators for inspection upon request.

➤ I-9 forms may be ordered in bulk from the Superintendent of Documents, U.S. Government Printing Office, Washington, DC 20402. Contractors may also call (202) 512-1800. Requests for information and questions about the implementation of the Immigration Reform and Control Act can be directed to (800) 375-5283.

➤ USCIS has additional information about the I-9 requirement on online at http://www.uscis.gov/i-9.

➤ An I-9 form must be completed for each newly hired employee. IRCA also prohibits discrimination. Under this law, contractors with four or more employees may not discriminate against any individual (other than an unauthorized alien) in hiring, termination, or recruiting or referring for a fee because of that individual's national origin or, in the case of a citizen or intending citizen, because of his or her citizenship status.

---

[4] For more information, please see OFCCP Directive Number: 284 - The U.S. Department of Homeland Security (DHS), U.S. Citizenship and Immigration Services' (USCIS) revised Employment Eligibility Verification Form (Form I-9).

➤ For more information concerning the anti-discrimination section of this law, contact the U.S. Department of Justice at:

U.S. Department of Justice
Civil Rights Division
Office of Special Counsel for Immigration-Related
Unfair Employment Practices
950 Pennsylvania Avenue, N.W.
Washington, D.C. 20530

Main Number: (202) 616-5594

Online: http://www.usdoj.gov/crt/osc.

Preparing for a Compliance Evaluation

The OFCCP conducts compliance evaluations to determine:

➢ Whether a contractor's affirmative action efforts comply with regulatory requirements;

➢ Whether a contractor has demonstrated good faith efforts in meeting its affirmative action requirements;

➢ Whether a contractor's employment policies and practices are free of discriminatory intent or impact;

➢ Whether a contractor has provided reasonable accommodation to qualified individuals with disabilities;

➢ Whether a contractor needs technical assistance to understand the evaluation process or to ensure that its affirmative action efforts are complete and effective; and

➢ How to best remedy any discriminatory practices or regulatory violations.

When contractors are notified of a compliance evaluation, they are given an overview of the procedures that OFCCP will use to conduct the evaluation. OFCCP compliance officers from regional and district offices will conduct the evaluation. Contractors should make sure that an officer of the company who is empowered to make and discuss policy and to make commitments for corrective action, where necessary, is present during the evaluation.

Contractors can prepare for a compliance evaluation by conducting a self-audit as a component of the affirmative action development process, or responding to inquiries likely to be asked by an OFCCP compliance officer during a evaluation. OFCCP compliance officers may ask to see documented evidence of a contractor's compliance efforts in the following areas:

Audit of Affirmative Action Specifications

➢ Does the contractor have written documentation of its efforts to comply with each of the 16 EEO and affirmative action specifications shown in the previous section of this Guide?

External Dissemination of Policy

➢ Has the contractor conspicuously displayed the required EEO poster (available from any OFCCP office) at each work site or company location in areas accessible to both applicants and employees?

➢ Do the contractor's contracts and purchase order forms display or reference the equal opportunity clauses as required?

Internal Dissemination of Policy

➢ At the start of each new job, has the contractor reviewed its EEO policy and affirmative obligations with all on-site supervisory and management personnel? Has the contractor kept records of these reviews?

➢ Per the guidelines on discrimination because of religion or national origin (41 CFR Part 60-50), have employees been informed of the contractor's commitment to equal employment opportunity for all persons, without regard to religion or national origin?

Community Relations

➢ Does the contractor use or solicit offers for subcontracts to minority and female owned businesses?

Audit of Personnel Operations

➢ Does the facility have written personnel policies and procedures? Have these policies and practices had an adverse impact on minorities, women, qualified individuals with disabilities, qualified disabled veterans, or other protected veterans?

➢ Are job descriptions in written form? Are job criteria objective and job-related?

Maintenance of Records

➢ Does the contractor maintain proper applicant flow records?

➢ Does the contractor maintain proper records about terminations and separations?

➢ Does the contractor maintain a system for identifying minority and female applicants and applicants who are individuals with disabilities and protected veterans for future consideration?

Validation

➢ Are written employment tests used by the contractor? If so, does the use of the test have an adverse impact on the hiring of minorities or women? Have tests been validated to ensure that they are valid predictors of an applicant's success in that position?

Directing Recruitment Efforts

➢ What recruitment sources are used by the contractor? Do these sources refer women, minorities, qualified individuals with disabilities, qualified disabled veterans, recently separated veterans, and other covered veterans?

Effect of Personnel Practices

➢ Are applicant processing procedures carried out in a uniform, nondiscriminatory fashion?

➢ Is there a disparity between the separation and termination rate of minorities and women as compared to non-minorities and males; or for individuals with disabilities and protected veterans as compared to individuals without disabilities and those who are not protected veterans? If so, why is that?

➢ Are there any restrictions to the granting of fringe benefits, including medical and life insurance, pension and retirement benefits, credit union benefits, and profit sharing and bonus plans based on the gender of the employee, status as a protected veteran or status as an individual with a disability?

➢ Are employment benefits available to the wives and families of male employees also available to the husbands and families of female employees? Are the benefits available to the families of individuals without disabilities and those who are not protected veterans also available to the families of individuals with disabilities and protected veterans?

➢ Does the contractor employ minorities, women, qualified individuals with disabilities, and protected veterans in each of its crafts? If so, to what extent? If not, what efforts has the contractor made to recruit members of these groups?

Training Programs

➢ Are training programs, including apprenticeship programs, available to employees without regard to race, color, sex, national origin, religion, or their status as an individual with a disability or protected veteran?

Compensation Disparities

➢ Do jobs offered by the contractor have similar duties but different pay rates? If so, do minorities or women earn less than their non-minority or male counterparts? Do individuals with disabilities or protected veterans earn less than their counterparts who are not individuals with disabilities or protected veterans?

➢ Do minorities, women, individuals with disabilities or protected veterans receive lower starting rates of pay than their counterparts with similar education and experience?

➢ Has the contractor reviewed its salary structure to ensure that it does not discriminate against minorities, women, individuals with disabilities or protected veterans?

Religion/National Origin

➢ Has the contractor reviewed its employment practices to determine whether members of various religious or ethnic groups receive fair consideration for job opportunities?

➢ Have reasonable accommodations to the religious observances and practices of employees or prospective employees been made, unless the accommodation would impose an undue hardship?

➢ Have recruiting sources been informed of the contractor's commitment to provide equal employment opportunity without regard to religion or national origin?

Sex Discrimination

➢ Does the contractor's policy on maternity leave meet regulatory requirements?

Harassment

➢ Has the contractor implemented policies and procedures to prevent, identify, and remedy instances of sexual harassment, and of harassment based on race, color, religion, national origin, gender, disability, or status as a protected veteran?

Retirement Policy

➢ Does the contractor's policy on mandatory or optional retirement age differ based upon the gender of the employee or their status as an individual with a disability or protected veteran?

Having answers and documentation for the above questions will go a long way towards preparing a contractor for an OFCCP compliance evaluation. Contractors should also know that when a compliance evaluation is scheduled, compliance officers will request the following documents for on-site inspection:

➢ Books, records, payrolls, accounts and other relevant documents, including a list, separated by construction project, of all employees who are members of protected groups who worked during the 12 months preceding this evaluation;

➢ Documentary evidence of the implementation of each of the specific affirmative action standards set forth in the sixteen specifications;

➢ A list of all Federal projects, including contract numbers, locations, estimated dollar values, percents completed and projected completion dates;

➢ A list of all non-Federal projects;

➢ A copy of the EEO-1 Report, where available;

➢ A copy of the written affirmative action program for individuals with disabilities, and the written affirmative action program for protected veterans (for contractors with Federal construction contracts only, not federally assisted construction contracts);

➢ A copy of the VETS-100 and VETS-100A reports (for contractors with Federal construction contracts only, not federally assisted construction contracts); and

➢ U.S. Citizenship and Immigration Service (USCIS) I-9 Forms.

Recognizing Good Faith Effort

Each year OFCCP hosts an award ceremony to recognize and honor those contractors and subcontractors that go well beyond the minimum requirements of the EEO and affirmative action laws.

The Secretary of Labor's **Opportunity Award**, initiated in 1988, is presented by the Secretary of Labor to honor one contractor for the successful implementation of a significant multi-faceted program ensuring equal employment opportunity and affirmative action within its organization and for the successful implementation of programs supporting these goals in the broader community.

The **Exemplary Voluntary Effort (EVE) Award**, initiated in 1983, is presented by the Deputy Assistant Secretary for Federal Contract Compliance to those contractors that have demonstrated through programs or activities, exemplary and innovative efforts to increase the employment opportunities for employees, including minorities, women, individuals with disabilities and covered veterans.

The **Exemplary Public Interest Contribution (EPIC) Award**, initiated in 1994, is presented by the Deputy Assistant Secretary for Federal Contract Compliance to honor selected public interest organizations that have supported affirmative action and linked their efforts with those of Federal contractors to enhance employment opportunities for minorities, women, individuals with disabilities and protected veterans.

The **G-FIVE Initiative,** initiated in 2008, recognizes companies' good faith efforts and best practices to employ and advance veterans. G-FIVE recognition is awarded by the Deputy Assistant Secretary for Federal Contract Compliance.

To be eligible for consideration for an EVE or Opportunity Award, a nominee must be a Federal contractor covered by Executive Order 11246, as amended; Section 503 of the Rehabilitation Act, as amended; and the Vietnam Era Veterans' Readjustment Assistance Act, as amended. Also, nominees must not have any unresolved violations of Federal law, as determined by compliance evaluations, complaint investigations, or Federal inspections and investigations.

In addition, the nominee must not have any enforcement actions pending, or be subject to any corrective actions or consent decrees that have resulted from litigation under laws enforced by the Department of Labor. While the EVE Award may be given for a single program or activity, recipients of the Opportunity 2000 Award must have developed and implemented a multi-faceted affirmative action program directed towards the changing demographics of the labor force. This may include involvement in community-based projects that assist in the development of a diverse workforce for the future. The Opportunity 2000 Award nominee may represent a single establishment or the entire corporation.

In past years, Opportunity and EVE Awards have recognized contractors' efforts that included:

➢ Recruitment, retention and management training and development programs that shatter glass ceilings and enhance opportunities for women and minorities at all levels of management;

➢ Innovative outreach and recruitment programs designed to attract minorities, women, qualified individuals with disabilities, and qualified protected veterans;

➢ Processes which provide individuals basic essential skills needed for employment;

➢ Programs that motivate and support minorities and women in attaining advanced degrees, and for education in science and technical fields;

➢ Seminars and conferences that create a greater awareness throughout a company of contributions of employees with learning disabilities; and

➢ Work place environment strategies that help employees balance work and family responsibilities.

To be eligible for consideration for an EPIC Award, a nominee must be a non-profit public interest organization whose activities support the mission of the OFCCP. Past winners have been recognized for their efforts in non-traditional employment for women, vocational training, literacy training, legal advocacy, scholarship programs, mentoring, and linkage with employment referrals to Federal contractors.

The latest guidance on the eligibility criteria, nomination process and administrative procedures for the Opportunity, EVE, and EPIC Awards can be found on OFCCP's web site at:

> http://www.dol.gov/ofccp/media/reports/eveint.htm

The latest guidance on the eligibility criteria, nomination process and administrative procedures for the G-FIVE Initiative can be found on OFCCP's web site at:

> http://www.dol.gov/ofccp/g_five.htm

Or contact your local OFCCP office for additional information regarding any of these honors.

Appendix A:

Glossary of Terms

(Adapted from the Federal Contract Compliance Manual)

Accessibility	The extent to which a contractor's facility is readily approachable and usable by individuals with disabilities, particularly such areas as the personnel office, job work sites, rest rooms and public areas.
Adverse impact	A substantially different rate of selection in hiring, promotion, transfer, training or other employment related decisions for any race, sex or ethnic group. *See definition of "Disparate impact."*
Affirmative Action	Actions, policies and procedures to which a contractor commits itself that are designed to achieve equal employment opportunity. Affirmative action obligations entail thorough, systematic efforts to prevent discrimination from occurring, to detect it and eliminate it as promptly as possible, and recruitment and outreach measures.
Applicant flow log	A chronological compilation of applicants for employment or promotion candidates, showing each individual categorized by race, sex and ethnic group, who applied for each job title (or group of job titles requiring similar qualifications) during a specific period.
Business necessity	A defense used by an employer when there is a selection criterion that is facially neutral but which excludes members of one sex, race, national origin or religious group at a substantially higher rate than members of other groups (thus creating an adverse impact). The employer must prove that the requirement having the adverse impact is job-related and consistent with business necessity. Business necessity may also have to be proven when a qualification standard screens out an individual because of their disability.
Civilian labor force	The aggregate of persons classified as employed and as unemployed in accordance with the criteria established by the Bureau of the Census and the U.S. Department of Commerce.
Compliance	Meeting the requirements and obligations imposed by Executive Order 11246, as amended, Section 503 of the

Rehabilitation Act of 1973, as amended, or 38 U.S.C. §4212, and their implementing regulations.

Construction contract   Any contract for the construction, rehabilitation, alteration, conversion, extension, demolition or repair of buildings or highways, or other changes or improvements to real property, including facilities providing utility services.

Construction site   The general physical location of any building, highway or real property undergoing construction, rehabilitation, alteration, conversion, extension, demolition, repair or any other change or improvement, and any other temporary location or facility at which a contractor or other participating party meets a demand or performs a function relating to the contract or subcontract.

Contract   Any Government contract or subcontract or any federally assisted construction contract or subcontract.

Contractor   A prime contractor or subcontractor, unless otherwise indicated.

Covered veteran   As used in this document, this term refers to any veteran who may be covered by 41 CFR Part 60-250 or 41 CFR Part 60-300. This term includes, but is not limited to, recently separated veterans, disabled veterans, Armed Forces service medal veterans, and veterans who served during a war or in a campaign or expedition for which a campaign badge has been authorized.

Disparate impact   A theory or category of employment discrimination. Disparate impact discrimination may be found when a contractor's use of a facially neutral selection standard (e.g., a test, an interview, a degree requirement) disqualifies members of a particular race, ethnic, or gender group at a significantly higher rate than others and is not justified by business necessity and job-relatedness. Intent to discriminate is not necessary to this type of employment discrimination.

Disparate treatment	A theory or category of employment discrimination. Disparate treatment discrimination may be found when a contractor treats an individual or group differently because of race, color, religion, sex, national origin, disability or veteran status. Intent to discriminate is a necessary element in this type of employment discrimination, and may be shown by direct evidence or inferentially by statistical, anecdotal and/or comparative evidence.
EEO-1 Report (or "Standard Form 100")	The Equal Employment Opportunity Employer Information Report. An annual report filed by certain employers subject to Executive Order 11246, as amended, or to Title VII of the Civil Rights Act of 1964, as amended. This report details the sex and race/ethnic composition of an employer's work force by job category. The EEO-1 Report is filed with the Joint Reporting Committee (JRC), which is composed of OFCCP and EEOC.
Employed	Under criteria established by the Bureau of the Census and the U.S. Department of Commerce, all civilians 16 years old and over who were either:

a)   "at work," meaning they performed at least some work during the reference week as paid employees or in their business or profession, or on their farm, or they worked 15 hours or more as unpaid workers on a family farm or in a family business; or

b)   "with a job but not at work," meaning they did not work during the reference week but had jobs or businesses from which they were temporarily absent due to illness, bad weather, industrial dispute, vacation, or other personal reasons.

Generally excluded from the category of "employed" are persons whose only activity consisted of unpaid work around the house or volunteer work for religious, charitable, and similar organizations, or persons on layoff.

Employee	A person employed by a Federal contractor, subcontractor or federally assisted construction contractor or subcontractor.
Establishment	A facility or unit which produces goods or services, such as a factory, office, store, or mine. In most instances, the unit is a physically separate facility at a single location. In

appropriate circumstances, OFCCP may consider as an establishment several facilities located at two or more sites when the facilities are in the same labor market or recruiting area. The determination as to whether it is appropriate to group facilities as a single establishment will be made by OFCCP on a case-by-case basis.

Facially neutral selection standards/criteria

A criterion/process is facially neutral if it does not make any reference to a prohibited factor and is equally applicable to everyone regardless of race, gender or ethnicity, *i.e.*, is not discriminatory on its face.

Federally assisted construction contract

Any agreement or modification thereof between any applicant and a person for construction work which is paid for in whole or in part with funds obtained from the Government or borrowed on the credit of the Government pursuant to any Federal program involving a grant, contract, loan, insurance or guarantee, or undertaken pursuant to any Federal program involving such grant, contract, loan, insurance or guarantee, or any application of modification thereof approved by the Government for a grant, contract, loan, insurance or guarantee under which the applicant itself participates in the construction work.

Immediate labor area

The geographic area from which employees reasonably may commute to the contractor's establishment. It may include one or more contiguous cities, counties, a metropolitan statistical area (MSA) or parts thereof.

Job area

Any sub-unit of a work force sector, such as a department, job group, job title, etc.

Job group

One or a group of jobs having similar content, wage rates and opportunities.

Labor area

Geographic area used in calculating availability. The area may vary from local to nationwide.

Non-compliance

A contractor's failure to adhere to the conditions set out in the contract's equal opportunity clauses and/or the regulations implementing those clauses and/or failure to correct violations.

Pattern or practice discrimination

Employer actions constituting a pattern of conduct resulting in discriminatory treatment toward the members of a class.

Prime contractor	Any person holding a contract subject to Executive Order 11246, as amended, Section 503 of the Rehabilitation Act of 1973, as amended, or VEVRAA, as amended; and for the purposes of 41 CFR 60-1, Subpart B; 41 CFR 60-250, Subpart B; 41 CFR 60-300, Subpart B; and 41 CFR 60-741, Subpart D, any person who has held a contract subject to Executive Order 11246, as amended, Section 503 of the Rehabilitation Act of 1973, as amended, or VEVRAA, as amended.
Protected veteran	*See definition of "Covered veteran."*
Reasonable accommodation	A reasonable accommodation can be:

a)   Modifications or adjustments to a job application process that enable a qualified individual (or veteran) with a disability to be considered for the position such applicant desires; or

b)   Modifications or adjustments to the work environment, or to the manner or circumstances under which the position held or desired is customarily performed, that enable a qualified individual (or veteran) with a disability to perform the essential functions of the position; or

c)   Modifications or adjustments that enable a contractor's employee with a disability to enjoy equal benefits and privileges of employment as are enjoyed by its other similarly situated employees without a disability.

d)   An employer does not have to provide any reasonable accommodation that will impose an undue hardship on its operations. *See definition of "Undue hardship."*

Subcontract	Any agreement or arrangement between a contractor and any person (in which the parties do not stand in the relationship of an employer and an employee):

a)   for the purchase, sale or use of supplies or services or the use of real or personal property, including lease arrangements, which, in whole or in part, is necessary to the performance of any one or more Government contracts; or

	b) under which any portion of the contractor's obligation under one or more Government contracts is performed, undertaken or assumed.
Subcontractor	Any person holding a subcontract, or anyone who has held a subcontract subject to Executive Order 11246, as amended, Section 503 of the Rehabilitation Act of 1973, as amended, or VEVRAA, as amended.
Systemic discrimination	Employment policies or practices that serve to differentiate or to perpetuate a differentiation in terms or conditions of employment of applicants or employees because of their status as members of a particular group, *e.g.*, a specific race or gender. Such policies may or may not be facially neutral, and intent to discriminate may or may not be involved.
Undue hardship	A defense used by an employer to explain why it did not provide a specific reasonable accommodation. The contractor must prove that providing the specific accommodation would have caused it significant difficulty or expense. Whether an accommodation would impose an undue hardship requires a case-by-case determination.
Validation	Validation is the demonstration of job-relatedness by showing the relationship between the selection procedure and job performance.
Veteran	*See definition of "Covered veteran."*
VETS-100 and VETS 100A Reports	The VETS-100 and VETS 100A Reports are to be completed by all nonexempt Federal contractors and subcontractors with contracts or subcontracts for the furnishing of supplies and services or for the use of real or personal property. VETS-100 must be completed by contractors with contracts entered into prior to December 1, 2003 for $25,000 or more. VETS-100A must be completed by contractors with contracts entered into or modified on or after December 1, 2003 for $100,000 or more. The Reports require that contractors report annually the numbers of various categories of veterans they employ or have newly hired by hiring location and job category.
Violation	Failure to fulfill a requirement of the Executive Order, Section 503, as amended, or VEVRAA, as amended, or their implementing rules, regulations and orders. The terms

"violation" and "deficiency" are often used interchangeably.

Appendix B:

Adverse Impact Determinations

Contractors with 100 or more employees must maintain and have available for each job records and other information showing the impact of the total selection process by identifiable race, sex and ethnic group. 41 CFR 60-3.4B and 3.15A(2)(a). "Total selection process" means the combined effect of all selection procedures leading to the final employment decision. At least annually, contractors with 100 or more employees are required to analyze these data to determine whether the total selection process for each job is having adverse impact. 41 CFR 60-3.15A(2). The adverse impact determinations must be conducted by gender and for each race or ethnic group (e.g., Black, Hispanic, Asian/Pacific Islander, and American Indian/Alaskan Native)[5] that constitutes 2 percent or more of the labor force in the relevant labor area or 2 percent or more of the applicable workforce. If the total selection process has an adverse impact, the impact of the individual components of the selection process also should be analyzed. 41 CFR 60-3.4C and 3.15A(2)(a).

"Adverse impact" is defined in the Uniform Guidelines as "a substantially different rate of selection in hiring, promotion, or other employment decision which works to the disadvantage of members of a race, gender, or ethnic group." 41 CFR 60-3.16B. Generally, to determine whether the differences in selection rates are sufficiently substantial to be regarded as evidence of adverse impact, the contractor should apply what is commonly referred to as the "4/5ths rule" or the "80 percent rule" of the Uniform Guidelines. Under this rule, a selection rate for any race, sex, or ethnic group that is less than 4/5ths or 80 percent of the selection rate for the group with the highest selection rate is generally regarded as evidence of adverse impact. The 80 percent rule is a general rule, and is not dispositive in all situations. The Uniform Guidelines recognize, for example, that sample size and other factors may affect the reliability of the 80 percent rule as a measure of adverse impact.

The 80 percent rule may not be accurate in detecting adverse impact where very large numbers of selections are made. Where the number of selections is very large, relatively small differences in selection rates may nevertheless constitute adverse impact if they are both statistically and practically significant. 41 CFR 60-3.4D. For that reason, where the sample size is very large, tests of practical and statistical significance should be used to assess whether the selection procedure results in adverse impact.

---

[5] OFCCP's regulations regarding the race, ethnicity, and job categories to be used by contractors have not changed to reflect the new categories for race, ethnicity, and job categories required for the EEO-1 Report. However, as a matter of enforcement discretion, OFCCP will not cite any contractor for non-compliance with the Executive Order solely because it utilizes the race, ethnicity, or job categories required by the new EEO-1 Report in records required by OFCCP regulations. Further, OFCCP will accept AAPs and supporting records that reflect the race, ethnicity, and job categories outlined in either 41 CFR Part 60-2 or the new EEO-1 Report. For more information, see OFCCP's Directive regarding the use of race and ethnic categories available online at: http://www.dol.gov/ofccp/regs/compliance/directives/dirindex.htm

Further, the 80 percent rule may not be a reliable indicator of adverse impact where the number of persons selected and difference in selection rates is very small. For example, if a contractor selected three males and one female from an applicant pool of 20 males and 10 females, the 80 percent rule would indicate adverse impact. The selection rate for women is 10 percent and the rate for men, 15 percent; 10/15 or 66 2/3 percent is less than 80 percent. Yet, the number of selections is too small to warrant a determination of adverse impact in these circumstances. Where the 80 percent rule indicates adverse impact, but the analysis is based on a sample too small to be reliable, evidence of the impact of the procedure over a longer period of time, or evidence concerning the impact of the procedure when used in the same manner elsewhere may be considered when determining adverse impact. 41 CFR 60-3.4D.

A four-step process is used to determine adverse impact:

1. Calculate the rate of selection for each group (divide the number of persons selected from a group by the number of applicants from that group).
2. Observe which group has the highest selection rate.
3. Calculate the impact ratios by comparing the selection rate for each group with that of the highest group (divide the selection rate for a group by selection rate for the highest group).
4. Observe whether the selection rate for any group is substantially less (*i.e.*, usually less than 4/5ths or 80 percent) than the selection rate for the highest group. If it is, adverse impact is indicated in most circumstances.

For example:

Applicants	Hires	Selection Rate
10 American Indians	2	2/10 or 20%
50 Blacks	20	20/50 or 40%
60 Hispanics	30	30/60 or 50%
80 Whites	48	48/80 or 60%

Comparisons of the selection rate for each group with that of the highest group (Whites) reveal the following impact ratios: American Indians 20/60 or 33%; Blacks 40/60 or 66.6%; and Hispanics 50/60 or 83%. Applying the 80 percent rule, on the basis of the above information, adverse impact is indicated for American Indians and Blacks but not for Hispanics.

If a selection procedure results in adverse impact, the contractor is required to eliminate it or justify its continued use. The contractor can justify using a selection procedure that has adverse impact by showing that the procedure has been validated according to the technical requirements of the Uniform Guidelines. "Validation" is the demonstration of job-relatedness by showing the relationship between the selection procedure and job performance. "Validation in accordance with the Guidelines" means a demonstration that a validity study meeting the standards of the Uniform Guidelines has been conducted and has produced evidence sufficient to warrant the use of the procedure for the purpose intended. 41 CFR 60-3.16X.

Even when a selection procedure with adverse impact has been validated, the contractor is obligated to investigate and consider suitable alternative selection procedures, and suitable alternative methods to using the selection procedure which have as little adverse impact as possible. 41 CFR 60-3.3B. Further, the contractor is required to use the procedure having less impact if it is "substantially equally valid." 41 CFR 60-3.3B.

There also are circumstances when a contractor may justify using a selection procedure with adverse impact by showing that it is required by "business necessity" *(i.e.,* the contractor must show that the selection procedure is job-related and necessary to the safe and efficient operation of its business).

In sum, the Uniform Guidelines recommend the following actions when adverse impact occurs:

- Modify the assessment instrument or procedure causing the adverse impact.

- Exclude the component procedure causing adverse impact from your selection process.

- Use an alternative procedure that causes little or no adverse impact, assuming that the alternative procedure is substantially equally valid.

- Use the selection procedure that has adverse impact only if the procedure is job-related and valid for predicting successful job performance, and there is no equally effective procedure available that has less adverse impact.

Appendix C:

The Small Business Administration's Ombudsman Program

The **Small Business Administration (SBA)**, in accordance with the provisions of the Small Business Regulatory Enforcement Fairness Act, has established a National Small Business and Agriculture Regulatory Ombudsman and 10 Regional Fairness Boards in order to receive comments from small businesses about federal regulatory enforcement actions. The SBA Ombudsman annually evaluates enforcement activities and rates each agency's responsiveness to small business. Small businesses wishing to comment on the enforcement activities of OFCCP may call 1-888-REG-FAIR (734-3247), or write to the SBA Ombudsman at:

SBA Ombudsman
409 Third Street SW
Washington, DC 20024
E-mail: ombudsman@sba.gov

Appendix D:

OFCCP National and Regional Offices

National Office

Office of Federal Contract Compliance Programs
Room C-3325
200 Constitution Avenue, NW
Washington, DC 20210
(202) 693-0101
(202) 693-1304 FAX

Northeast Region
(Connecticut, Maine, Massachusetts, New Hampshire New Jersey, New York, Puerto Rico, Rhode Island, Vermont, Virgin Islands)

Regional Office
201 Varick Street, Room 750
New York, NY 10014
(646) 264-3170
(646) 264-3009 FAX

Mid-Atlantic Region
(Delaware, District of Columbia, Maryland, Pennsylvania, Virginia, West Virginia)

Regional Office
Curtis Center, Suite 750 West
170 S. Independence Mall West
Philadelphia, PA 19106
(215) 861-5765
(215) 861-5769 FAX

Southeast Region
(Alabama, Florida, Georgia, Kentucky, Mississippi, North Carolina, South Carolina, Tennessee)

Regional Office
61 Forsyth Street, Room 7B75
Atlanta, GA 30303
(404) 893-4545
(404) 893-4546 FAX

Midwest Region
(Illinois, Indiana, Iowa, Kansas, Michigan, Minnesota, Missouri, Nebraska, Ohio, Wisconsin)

Regional Office
Kluczynski Federal Building, Room 570
230 South Dearborn Street
Chicago, IL 60604
(312) 596-7010
(312) 596-7037 FAX

Southwest and Rocky Mountain Region
(Arkansas, Colorado, Louisiana, Montana, New Mexico, North Dakota, Oklahoma, South Dakota, Texas, Utah, Wyoming)

Regional Office
525 South Griffin Street
Federal Building, Room 840
Dallas, TX 75202
(972) 850-2550
(972) 850-2552 FAX

Pacific Region
(Alaska, Arizona, California, Guam, Hawaii, Idaho, Nevada, Oregon, Washington)

Regional Office
90 7th Street, Suite # 18-300
San Francisco, CA 94103
(415) 625-7800
(415) 625-7799 FAX

Appendix E:

Participation Goals
for Minorities and Females

For federal and federally assisted construction contractors, goals for minorities and females are established as a percentage participation rate. The percentage goal established for minority participation must be at least equal to the percentage established for that "economic area" as outlined in the list below.[6]

Contractors may establish higher goals if they desire. Although a contractor is required to make good faith efforts to meet their goals, the goals are not quotas and no sanctions are imposed solely for failure to meet them. The following factors explain the difference between permissible goals, on the one hand, and unlawful preferences, on the other:

♦ Participation rate goals are not designed to be, nor may they properly or lawfully be interpreted as, permitting unlawful preferential treatment and quotas with respect to persons of any race, color, religion, sex, or national origin.

♦ Goals are neither quotas, set-asides, nor a device to achieve proportional representation or equal results. Rather, the goal-setting process is used to target and measure the effectiveness of affirmative action efforts to eradicate and prevent barriers to equal employment opportunity.

♦ Goals under Executive Order 11246, as amended, do not require that any specific position be filled by a person of a particular gender, race, or ethnicity. Instead, the requirement is that contractors engage in outreach and other efforts to broaden the pool of qualified candidates to include minorities and women.

♦ The use of goals is consistent with principles of merit, because goals do not require an employer to hire a person who does not have the qualifications needed to perform the job successfully, hire an unqualified person in preference to another applicant who is qualified, or hire a less qualified person in preference to a more qualified person.

♦ Goals may not be treated as a ceiling or a floor for the employment of members of particular groups.

♦ A contractor's compliance is measured by whether it has made good faith efforts to meet its goals, and failure to meet goals, by itself, is not a violation of the Executive Order.

These goals are applicable to all of a contractor's construction work sites (whether or not these sites are also the result of a federal contract or are federally assisted). The goals are applicable to

---

[6] For more information about the development of the goals, see Federal Register, Vol. 45, No. 194, at 65976-65991 (October 3, 1980) (minorities) and Federal Register, Vol. 45, No. 251 at 85750-85751 (December 30, 1980) (females). The text of these Federal Register notices can be found:

- Federal Register Notice : Vol. 45, No. 194, at 65976-65991 (October 3, 1980)   [HTML] | [PDF]
- Federal Register Notice : Vol. 45, No. 251, at 85750-85751 (December 30, 1980) [HTML] | [PDF]

each nonexempt contractor's total onsite construction workforce, regardless of whether or not part of that workforce is performing work on a federal, federally assisted or non-federally related project contract or subcontract.

Contractors should apply to each work site the goal for the geographical area that each particular work site is located in.

The contractor's compliance with the Executive Order and the regulations in 41 CFR Part 60-4 will be assessed based on its implementation of the Equal Opportunity Clause, specific affirmative action obligations required by the specifications set forth in 41 CFR 60-4.3(a), and its efforts to meet the goals. The hours of minority and female employment and training must be substantially uniform throughout the length of the contract, and in each trade, and the contractor must make a good faith effort to employ minorities and women evenly on each of its projects. The transfer of minority or female employees or trainees from contractor to contractor or from project to project for the sole purpose of meeting the contractor's goals is a violation of the contract, the Executive Order, and the regulations in 41 CFR Part 60-4. Compliance with the goals will be measured against the total work hours performed.

Until further notice, the following goals for female and minority utilization in each construction craft and trade must be included in all Federal or federally assisted construction contracts and subcontracts in excess of $10,000 .

Construction contractors that are participating in an approved Hometown Plan (see 41 CFR 60-4.5) are required to comply with the goals of the Hometown Plan with regard to construction work they perform in the area covered by the Hometown Plan. With regard to all their other covered construction work, such contractors are required to comply with the applicable SMSA or EA goal contained in the list below.

GOALS FOR FEMALES[7]

Nationwide Goal_____**6.9%**

---

[7] The percentage goal established for female participation is **6.9%** nationwide.

GOALS FOR MINORITIES

ECONOMIC AREAS

STATE	GOAL (percent)

Maine:

001 Bangor, ME:
  Non-SMSA Counties _____ 0.8
    ME Aroostook; ME Hancock; ME Penobscot; ME Piscataquis; ME Waldo; ME Washington.
002 Portland-Lewiston, ME:
  SMSA Counties:
    4243 Lewiston-Auburn, ME _____ 0.5
      ME Androscoggin.
    6403 Portland, ME   0.6
      ME Cumberland; ME Sagadahoc.
  Non-SMSA Counties _____ 0.5
    ME Franklin; ME. Kennebec; ME Knox. ME; Lincoln; ME Oxford; ME Somerset; ME York.

Vermont:

003 Burlington, VT:
  Non-SMSA Counties _____ 0.8
    NH Coos; NH Grafton; NH Sullivan; VT Addison; VT Caledonia; VT Chittenden; VT Essex; VT
    Franklin; VT Grand Isle; VT Lamoille; VT Orange; VT Orleans; VT Ruthland; VT Washington; VT
    Windsor.

Massachusetts:

004 Boston, MA:
  SMSA Counties:
    1123 Boston - Lowell - Brockton - Lawrence - Haverhill. MA-NH _____ 4.0
      MA Essex; MA Middlesex; MA Norfolk; MA Plymouth; MA Suffolk; NH Rockingham.
    4763 Manchester-Nashua, NH _____ 0.7
      NH Hillsborough.
    5403 Fail River-New Bedford, MA _____ 1.6
      MA Bristol
    9243 Worcester - Fitchburg – Leominster, MA _____ 1.6
      MA Worcester.
  Non-SMSA Counties _____ 3.6
    MA Barnstable; MA Dukes-, MA Nantucket, NH Belknap; NH Carroll; NH Merrimack; NH
    Strafford.

Rhode Island:

005 Providence - Warwick - Pawtucket, RI:
  SMSA Counties:
    6483 Providence - Warwick - Pawtucket RI _____ 3.0
      RI Bristol; RI Kent; RI Providence; RI Washington
  Non-SMSA Counties _____ 3.1
    RI Newport.

Connecticut (Massachusets):

006 Hartford - New Haven - Springfield, CT-MA:
  SMSA Counties:
    3283 Hartford - New Britain – Bristol, CT _____ 6.9
      CT Hartford; CT Middlesex; CT Tolland
    5483 New Haven - Waterbury – Meriden, CT _____ 9.0
      CT New Haven.
    5523 New London - Norwich, CT _____ 4.5
      CT New London.
    6323 Pittsfield, MA _____ 1.6
      MA Berkshire.
    8003 Springfield - Chicopee - Holyoke. MA-CT _____ 4.8
      MA Hampden; MA Hampshire.
  Non-SMSA Counties _____ 5.9
    CT Litchfield; CT Windham; MA Franklin; NH Cheshire; VT Windham.

New York:

007 Albany - Schenectady - Troy, NY:
  SMSA Counties:
    0160 Albany - Schenectady – Troy, NY _____ 3.2
    NY Albany; NY Montgommy, NY Rensselaer, NY Saratoga; NY Schenectady.
  Non-SMSA Counties _____ 2.6
    NY Clinton; NY Columbia; NY Essex; NY Fulton; NY Greene; NY Hamilton, NY Scho-
    harie, NY Warren; NY Washington; VT Bennington.
008 Syracuse - Utica, NY:
  SMSA Counties:
    8160 Syracuse _____ 3.8
    NY Madison; NY Onondaga; NY Oswego.
    8680 Utica -Rome, NY _____ 2.1
    NY Herkimer; NY Oneida.
  Non-SMSA Counties _____ 2.5
    NY Cayuga; NY Cortland; NY Franklin; NY Jefferson; NY Lewis; NY St. Lawrence.
009 Rochester, NY:
  SMSA Counties:
    6840 Rochester, NY _____ 5.3
    NY Livingston; NY Monroe; NY Ontario; NY Orleans; NY Wayne.
  Non-SMSA Counties _____ 5.9
    NY Genesee; NY Seneca; NY Yates.
010 Buffalo, NY:
  SMSA Counties:
    1280 Buffalo, NY _____ 7.7
    NY Erie; NY Niagara.
  Non-SMSA Counties _____ 6.3
    NY Allegany; NY Cattaraugus; NY Chautauqua; NY Wyoming. PA McKean; PA Potter.
011 Binghamton - Elmira, NY:
  SMSA Counties
    0960 Binghamton, NY - PA _____ 1.1
    NY Broome; NY Tioga; PA Susquehanna.
    2335 Elmira, NY _____ 2.2
    NY Chemung
  Non-SMSA Counties _____ 1.2
    NY Chenango; NY Delaware; NY Otsego; NY Schuyler; NY Steuben; ; NY Tompkins; PA
    Bradford; PA Tioga.

012 New York, NY:
   SMSA Counties:
      1163 Bridgeport - Stamford - Norwalk - Danbury, CT_____10.2
      CT Fairfield.
      3640 Jersey City, NJ_____12.8
      NJ Hudson.
      4410 Long Branch - Asbury Park, NJ_____9.5
      NJ Monmouth
      5380 Nassau – Suffolk, NY_____5.8
      NY Nassau; NY Suffolk.
      5460 New Brunswick - Perth Amboy - Sayreville, NJ.
      NJ Middlesex _____5.8
      5600 New York NY, NJ
      NJ Bergen; NY Putnam; NY Rockland; NY Westchester_____22.6

The following goal ranges are applicable to the indicated trades in the counties of Bronx, Kings, New York, Queens, and Richmond.

Trade	Goal range
Electricians	9.0 to 10.2
Carpenters	27.6 to 32.0
Steam fitters	12.2 to 13.5
Metal lathers	24.6 to 25.6
Painters	28.6 to 26.0
Operating Engineers	25.6 to 26.0
Plumbers	12.0 to 14.5
Iron workers (struct)	25.9 to 32.0
Elevator constructors	5.5 to 6.5
Bricklayers	13.4 to 15.5
Asbestos workers	22.8 to 28.0
Roofers	6.3 to 7.5
Iron workers (ornamental)	22.4 to 23.0
Cement masons	23.0 to 27.0
Glaziers	16.0 to 20.0
Plasterers	15.8 to 18.0
Teamsters	22.0 to 22.5
Boilermakers	13.0 to 15.5
All others	16.4 to 17.5

      5640 Newark, NJ_____17.3
      NJ Essex; NJ Morris; NJ Somerset; NJ Union.
      6040 Paterson - Clifton - Passaic., NJ_____12.9
      NJ Passaic.
      6460 Poughkeepsie, NY_____6.4
      NY Dutchese
    Non-SMSA Counties_____17.0
      NJ Hunterdon; NJ Ocean; NJ Sussex; NY Orange; NY Sullivan; NY Ulster; PA Pike.

Pennsylvania:

013 Scranton - Wilkes-Barre, PA:
   SMSA Counties
      5745 Northeast Pennsylvania_____0.6
      PA Lackawanna; PA Luzerne; PA Monroe.
   Non-SMSA Counties   0.5
      PA Columbia; PA Wayne; PA Wyoming.
014 Williamsport, PA:
   SMSA Counties
      9140 Williamsport, PA_____1.0

PA Lycoming.
Non-SMSA Counties   0.7
PA Cameron; PA Centre; PA Clearfield; PA Clinton; PA Elk; PA Jefferson; PA Montour;
PA Northumberland; PA Snyder; PA Sullivan; PA Union.

015 Erie, PA:
  SMSA Counties:
    2360 Erie, PA_____2.8
      PA Erie.
  Non-SMSA Counties_____1.8
    PA Clarion; PA Crawford; PA Forest; PA Venango; PA Warren.

016 Pittsburgh, PA:
  SMSA Counties
    0280 Altoona, PA_____1.0
      PA Blair.
    3680 Johnson, PA_____1.3
      PA Cambria; PA Somerset.
    6280 Pittsburgh, PA_____6.3
      PA Allegheny; PA Beaver; PA Washington; PA Westmoreland.
  Non-SMSA Counties_____4.8
    MD Allegany; MD Garrett; PA Armstrong; PA Bedford; PA Butler; PA Fayette; PA
    Greene; PA Indiana; WV Mineral.

017 Harrisburg - York - Lancaster, PA:
  SMSA Counties
    3240 Harrisburg PA_____6.2
      PA Cumberland; PA Dauphin; PA Perry.
    4000 Lancaster, PA_____2.0
      PA Lancaster.
    9280 York, PA_____2.2
      PA Adams; PA York.
  Non-SMSA-Counties_____3.1
    PA Franklin; PA Fulton; PA Huntingdon; PA Juniata; PA Lebanon; PA Mifflin.

018 Philadelphia, PA:
  SMSA Counties
    0240 Allentown - Bethlehem - Easton, PA-NJ_____1.6
      NJ Warren; PA Carbon; PA Lehigh; PA Northhampton.
    0560 Atlantic City, NJ_____18.2
      NJ Atlantic
    6160 Philadelphia, PA-NJ_____17.3
      NJ Burlington; NJ Camden; NJ Glouchester; PA Bucks; PA Chester; PA Delaware; PA
      Montgomery; PA Philadelphia.
    8680 Reading, PA_____2.5
      PA Berks.
    8480 Trenton, NJ_____16.4
      NJ Mercer.
    8760 Vineland - Millville - Bridgeton, NJ_____16.0
      NJ Cumberland.
    9160 Wilmington, DE-NJ-MD_____12.3
      DE Now Castle; MD Cecil; NJ Salem.
  Non-SMSA Counties_____14.5
    DE Kent; DE Sussex; NJ Cape May; PA Schuylkill.

Maryland:

019 Baltimore, MD
SMSA Counties:
    0720 Baltimore MD_____23.0
      MD Anne Arundel; MD Baltimore; MD Carroll; MD Harford; MD Howard; MD Baltimore City.
Non-SMSA Counties_____23.6

MD Caroline; MD Dorchester; MD Kent; MD Queen Annes; MD Somerset; MD Talbot; MD Wicomico; MD Worcester; VA Accomack; VA Northamptom

Washington DC:

020 Washington DC:
SMSA Counties
    8840 Washington, DC-MD-VA_____28.0
    DC District of Columbia; MD Charles; Montgomery; MD Prince Georges; VA Arlington; VA Fairfax; VA Loudoun; VA Prince William; VA Alexandria; VA Fairfax City; VA Falls Church.
Non-SMSA Counties_____25.2
    MD Calvert; MD Frederick; MD St. Marys; MD Washington; VA Clarke; VA Culpepper; VA Fauquier; VA Frederick; VA King George; VA Page; VA Rappahannock; VA Shenandoah; VA Spottsylvania; VA Stafford; VA Warren; VA Westmoreland; VA Fredericksburg; VA Winchester; WV Berkeley; WV Grant; WV Hampshire; WV Hardy; WV Jefferson; WV Morgan.

Virginia:

021 Roanoke-Lynchburg VA:
SMSA Counties:
    4640 Lynchburg, VA_____19.3
    VA Amherst; VA Appomattox; VA Campbell; VA Lynchburg.
    6800 Roanoke, VA_____10.2
    VA Botetourt; VA Craig; VA Roanoke VA; VA Roanoke City; VA Salem.
Non-SMSA Counties_____12.0
    VA Alleghany, VA Augusta; VA Bath; VA Bedford; VA Bland; VA Carroll; VA Floyd; VA Franklin; VA Giles; VA Grayson; VA Henry; VA Highland; VA Montgomery; VA Nelson; VA Patrick; VA Pittsylvania; VA Pulaski; VA Rockbridge; VA Rockingham; VA Wythe; VA Bedford City; VA Buena Vista; VA Clifton Forge; VA Covington; VA Danville; VA Galex; VA Harrisonburg; VA Lexington; VA Martinsville; VA Radford; VA Staunton; VA Waynesboro; WV Pendleton.
022 Richmond, VA:
SMSA Counties:
    6140 Petersburg - Colonial Heights – Hopewell, VA_____30.6
    VA Dinwiddie; VA Prince George; VA Colonial Heights; VA Hopewell; VA Petersburg.
    6760 Richmond, VA_____24.9
    VA Charles City; VA Chesterfield; VA Goochland; VA Hanover; VA Henrico; VA New Kent; VA Powhatan; VA Richmond.
Non-SMSA Counties_____27.9
    VA Albemarle; VA Amelia; VA Brunswick; VA Buckingham; VA Caroline; VA Charlotte; VA Cumberland; VA Essex; VA Fluvanna; VA Greene; VA Greensville; VA Halifax; VA King And Queen; VA King William; VA Lancaster; VA Louisa; VA Lunenberg; VA Madison; VA Mecklenburg; VA Northumberland; VA Nottoway; VA Orange; VA Prince Edward; VA Richmond; VA Sussex; VA Charlottesville; VA Emporia; VA South Boston.
023 Norfolk - Virginia Beach - Newport News, VA:
SMSA Counties:
    5680 Newport News-Hampton, VA_____27.1
    VA Gloucester, VA James City; VA York; VA Hampton; VA Newport News; VA Williamsburg.
    5720 Norfolk - Virginia Beach – Portsmouth VA – NC_____26.6
    NC Currituck; VA Chesapeake; VA Norfolk; VA Portsmouth; VA Suffolk; VA Virginia Beach.

Non-SMSA Counties_____29.7
    NC Bertie; NC Camden; NC Chowan; NC Gates; NC Hertford; NC Pasquotank; NC Perquimans; VA Isle of Wight; VA Matthews; VA Middlesex; VA Southampton; VA Surry; VA Franklin.

North Carolina:

024 Rocky Mount - Wilson - Greenville NC:
  Non-SMSA Counties_____31.7
    NC Beaufort; NC Carteret; NC Craven; NC Dare; NC Edgecombe; NC Greene; NC
    Halifax; NC Hyde; NC Jones; NC Lenoir; NC Martin; NC Nash; NC Northampton; NC
    Pamlico; NC Pitt; NC Tyrrell; NC Washington; NC Wayne; NC Wilson
025 Wilmington, NC:
  SMSA Counties:
    9200 Wilmington, NC_____20.7
    NC Brunswick; NC New Hanover.
  Non-SMSA counties_____23.5
    NC Columbus; NC Duplin; NC Onslow; NC Pender.
026 Fayetteville, NC:
  SMSA Counties:
    2560 Fayetteville, NC_____26.2
    NC Cumberland.
  Non-SMSA Counties_____33.5
    NC Bladen; NC Hoke; NC Richmond; NC Robeson; NC Sampson; NC Scotland.
027 Raleigh - Durham, NC.
  SMSA Counties:
    6640 Raleigh – Durham_____22.8
    NC Durham; NC Orange; NC Wake.
  Non-SMSA Counties_____24.7
    NC Chatham; NC Franklin; NC Granville; NC Harnett; NC Johnston; NC Lee; NC Person;
    NC Vance; NC Warren.
028 Greensboro - Winston Salem - High Point, NC:
  SMSA Counties:
    1300 Burlington, NC_____16.2
    NC Alamance.
    3120 Greensboro - Winston Salem – High Point NC_____16.4
    NC Davidson; NC Forsyth; NC Guilford; NC Randolf; NC Stokes; NC Yadkin.
  Non-SMSA Counties_____15.5
    NC Alleghany; NC Ashe; NC Caswell; NC Davie; NC Montgomery; NC Moore; NC
    Rockingham; NC Surry; NC Watauga; NC Wilkes.
029 Charlotte, NC:
  SMSA Counties:
    1520 Charlotte – Gastonia, NC_____18.5
    NC Gaston; NC Mecklenburg; NC Union.
  Non-SMSA Counties_____15.7
    NC Alexander; NC Anson; NC Burke; NC Cabarrus; NC Caldwell; NC Catawba;
    NC Cleveland; NC Iredell; NC Lincoln; NC Rowan; NC Rutherford; NC Stanley;
    SC Chester; SC Lancaster SC York.
030 Asheville, NC
  Non-SMSA Counties:
    0480 Asheville, NC_____8.5
    NC Buncombe; NC Madison.
  Non-SMSA Counties_____6.3
    NC Avery; NC Cherokee; NC Clay; NC Graham; HC Heywood; NC Henderson;
    NC Jackson; NC McDowell; NC Macon; NC Mitchell; NC Swain; NC Transylvania;
    NC Yancey.

South Carolina:

031 Greenville – Spartanburg, SC:
  SMSA Counties:
    3160 Greenville –Spartanburg, SC_____16.0
    SC Greenville; SC Pickens; SC Spartanburg.
  Non-SMSA Counties_____17.8

SC Polk; SC Abbeville; SC Anderson; SC Cherokee; SC Greenwood; SC Laurens; SC Oconee; SC Union.

032 Columbia, SC
  SMSA Counties:
    1760 Columbia, SC_____23.4
    SC Lexington; SC Richland.
  Non-SMSA Counties_____32.0
    SC Calhoun SC Clarendon; SC Fairfield; SC Kershaw; SC Lee; SC Newberry; SC Orangeburg; SC Saluda; SC Sumter

033 Florence, SC
  Non-SMSA Counties_____33.0
    SC Chesterfield; SC Darlington; SC Dillon; SC Florence; SC Georgetown; SC Horry; SC Marion; SC Marlboro; SC Williamsburg.

034 Charleston - North Charleston, SC
  SMSA Counties
    1440 Charleston - North Charleston, SC_____30.0
    SC Berkeley; SC Charleston; SC Dorchester.
  Non-SMSA Counties_____30.7
    SC Collection

Georgia:

035 Augusta, GA:
  SMSA Counties:
    0600 Augusta, GA – SC_____27.2
    GA Columbia; GA Richmond; SC Aiken
  Non-SMSA Counties_____32.8
    GA Burke; GA Emanuel; GA Glascock; GA Jefferson; GA Jenkins; GA Lincoln; GA McDuffie; GA Taliaferro; GA Warren; GA Wilkes; SC Allendale, SC Bamberg; SC Barnwell; SC Edgefield; SC McCormick

036 Atlanta, GA
  SMSA Counties
    0520 Atlanta_____21.2
    GA Butts; GA Cherokee; GA Clayton; GA Cobb; GA Dekalb; GA Douglas; GA Fayette; GA Forsyth; GA Fulton; GA Gwinnett; GA Henry, GA Newton; GA Paulding; GA Rockdale; GA Walton
  Non-SMSA Counties_____19.5
    GA Banks; GA Barrow; GA Bartow; GA Carroll; GA Clarke; GA Coweta; GA Dawson; GA Elbert; GA Fannin; GA Floyd; GA Franklin; GA Gilmer; GA Gordon; GA Greene; GA Habersham; GA Hall; GA Haralson; GA Hart; GA Heard; GA Jackson; GA Jasper; GA Lamar, GA Lumpkin; GA Madison, GA Morgan; GA Oconee, GA Oglethorpe; GA Pickens; GA Pike; GA Polk; GA Rabun, GA Spalding; GA Stephens; GA Towns; GA Union; GA Upson; GA White.

037 Columbus, GA:
  SMSA Counties
    1800 Columbus_____29.6
    AL Russell; GA Chattahoochee; GA Columbus.
  Non-SMSA Counties_____31.6
    AL Chambers; AL Lee; GA Harris; GA Marion; GA Meriwether; GA Quitman; GA Schley; GA Stewart; GA Sumter; GA Talbot; GA Troup; GA Webster.

---

038 Macon, GA:
  SMSA Counties
    4660 Macon, GA _____27.5
      GA Bibb; GA Houston; GA Jones; GA Twiggs.
  Non-SMSA Counties_____31.7
      GA Baldwin; GA Bleckley; GA Crawford; GA Crisp; GA Dodge; GA Dooly; GA Hancock;
      GA Johnson; GA Laurens; GA Macon; GA Monroe; GA Peach; GA Pulaski; GA
      Putnam. GA Taylor; GA Telfair; GA Treutlen; GA Washington; GA Wheeler; GA Wilcox;
      GA Wilkinson.
039 Savannah, GA:
  SMSA Counties:
    7520 Savannah, GA_____30.6
      GA Bryan; GA Chatham; GA Effingham
  Non-SMSA Counties_____29.8
      GA Appling; GA Atkinson; GA Bacon; GA Bullock; GA Candler; GA Coffee;
      GA Evans; GA Jeff Davis; GA Liberty; GA Long; GA McIntosh; GA Montgomery;
      GA Screven; GA Tattinall; GA Toombs; GA Wayne; SC Beaufort; SC Hampton; SC Jasper.
040 Albany, GA
  SMSA Counties
    0120 Albany, GA_____32.1
      GA Dougherty; GA Lee.
  Non-SMSA Counties_____31.1
      GA Baker; GA Ben Hill; GA Berrien; GA Brooks; GA Calhoun; GA Clay; GA Clinch; GA
      Colquitt; GA Cook; GA Decatur; GA Early; GA Echols; GA Grady; GA Irwin; GA Lanier;
      GA Lowndes; GA Miller; GA Mitchell; GA Randolph; GA Seminole, GA Terrell; GA
      Thomas; GA Tift; GA Turner; GA Worth.

Florida:

041 Jacksonville, FL:
  SMSA Counties
    2900 Gainesville, FL_____20.6
      FL Alachua
    3600 Jacksonville, FL_____21.8
      FL Baker; FL Clay; FL Duval; FL Nassau; FL St. Johns.
  Non-SMSA Counties_____22.2
      FL Bradford; FL Columbia; FL Dade; FL Gilchrist; FL Hamilton; FL LaFayetle;
      FL Levy; FL Marion; FL Putnam; FL Suwannee; FL Union; GA Brantley; GA Camden;
      GA Charlton; GA Glynn; GA Pierce; GA Ware.
042 Orlando - Melbourne - Daytona Beach, FL.
  SMSA Counties:
    2020 Daytona Beach, FL_____15.7
      FL Volusia.
    4900 Melbourne – Tutusville – Cocoa, FL _____10.7
      FL Brevard.
    5960 Orlando, FL_____15.5
      FL Orange; FL Osceola; FL Seminole.
  Non-SMSA Counties_____14.9
      FL Flagler; FL Lake; FL Sumter.
043 Miami - Fort Lauderdale, FL:
  SMSA Counties:
    2680 Fort Lauderdale – Hollywood, FL_____15.5
      FL. Broward.
    5000 Miami, FL_____39.5
      FL Dade.
    8960 West Palm Beach - Boca Raton, FL_____22.4
      FL Palm Beach.

Non-SMSA Counties_____30.4
    FL Glades; FL Hendry; FL Indian River; FL Martin; FL Monroe;
    FL Okeechobee; FL St. Lucie.
044 Tampa - St Petersburg, FL
  SMSA Counties:
    1140 Bradenton, FL_____15.9
    FL Manatee.
    2700 Fort Myers, FL_____15.3
    FL Lee.
    3980 Lakeland - Winter Haven, FL_____18.0
    FL Polk
    7510 Sarasota, FL_____10.5
    FL Sarasota.
    8280 Tampa - St. Petersburg, FL_____17.9
    FL Hillsborough, FL Pasco; FL Pinellas
  Non-SMSA Counties_____17.1
    FL Charlotte; FL Citrus; FL Collier, FL Desoto; FL Hardee; FL Hernando; FL Highlands.
045 Tallahassee. FL:
  SMSA Counties:
    8240 Tallahassee, FL_____24.3
    FL Leon; FL Wakulla.
  Non-SMSA Counties:_____29.5
    FL Calhoun; FL Franklin; FL Gadsden; FL Jackson; FL Jefferson; FL Liberty;
    FL Madison; FL Taylor.
046 Pensacola - Panama City, FL
  SMSA Counties:
    8615 Panama City, FL_____14.1
    FL Bay.
    6080 Pensacola, FL_____18.3
    FL Escambia; FL Santa Rosa.
  Non-SMSA Counties_____15.4
    FL Gulf; FL Holmes; FL Okaloosa; FL Walton; FL Washington.

Alabama:

047 Mobile, AL
  SMSA Counties:
    5160 Mobile, AL_____25.9
    AL Baldwin; AL Mobile.
    6026 Pascagoula - Moss, Point MS_____16.9
    MS Jackson.
  Non-SMSA Counties_____26.4
    AL Choctaw; AL Clarke; AL Conecuh; AL Escambia; AL Monroe; AL Washington; AL Wilcox; MS
    George; MS Greene.
048 Montgomery, AL:
  SMSA Counties
    5240 Montgomery, AL_____29.9
    AL Autauga; AL Elmore; AL Montgomery.
  Non-SMSA Counties_____29.9
    AL Barbour; AL Bullock; AL Butler; AL Coffee; AL Coosa; AL Covington;
    AL Crenshaw; AL Dale; AL Dallas; AL Geneva; AL Henry; AL Houston;
    AL Lowndes; AL Macon; AL Perry; AL Pike; AL Tallapoosa.
049 Birmingham, AL:
  SMSA Counties:
    0450 Anniston, AL_____14.3
    AL Calhoun.

1000 Birmingham, AL_____24.9
  AL Jefferson; AL St. Clair; AL Shelby; AL Walker; AL Etowah
8600 Tuscaloosa, AL_____20.6
  AL Tuscaloosa.
Non-SMSA Counties_____20.7
  AL Bibb; AL Blount; AL Cherokee; AL Chilton; AL Clay; AL Cleburne; AL Cullman;
  AL Fayette; AL Greene; AL Hale; AL Lamar; AL Marion; AL Pickens; AL Randolph;
  AL Sumter; AL Talladega; AL Winston.
050 Huntsville – Florence, AL:
  SMSA Counties:
    2650 Florence, AL_____11.9
      AL Colbert; AL Lauderdale.
    3440 Huntsville, AL_____12.0
      AL Limestone; AL Madison; AL Marshall.
    Non-SMSA Counties_____11.2
      AL Franklin; AL Lawrence AL Morgan; TN Lincoln.

Tennessee:

051 Chattanooga, TN:
  SMSA Counties:
    1560 Chattanooga, TN – GA_____12.5
      GA Catoosa; GA Dade; GA Walker; TN Hamilton;TN Marion; TN Sequatchie.
    Non-SMSA Counties_____8.6
      AL De Kalb; AL Jackson; GA Chattooga; GA Murray; GA Whitfield;
      TN Bledsoe; TN Bradley; TN Grundy; TN McMinn; TN Meigs; TN Monroe;
      TN Polk; TN Rhea.
052 Johnson City - Kingsport - Bristol, TN-VA:
  SMSA Counties:
    3660 Johnson City - Kingsport - Bristol. TN – VA_____2.6
      TN Carter; TN Hawkins; TN Sullivan; TN Unicoi; TN Washington; VA Scott;
      VA Washington; VA Bristol.
    Non-SMSA Counties_____3.2
      TN Greene; TN Hancock; TN Johnson; VA Buchanan; VA Dickenson; VA Lee;
      VA Russell; VA Smyth; VA Tazewell; VA Wise; VA Norton; WV McDowell, WV Mercer.
053 Knoxville, TN
  SMSA Counties:
    3840 Knoxville, TN_____6.6
      TN Anderson; TN Blount; TN Knox; TN Union.
    Non-SMSA Counties_____4.5
      KY Bell; KY Harlan; KY Knox; KY Laurel; KY McCreary; KY Wayne; KY Whitley; TN
      Campbell; TN Claiborne; TN Cocke; TN Cumberland; TN Fentress; TN Grainger,
      TN Hamblen; TN Jefferson; TN Loudon; TN Morgan; TN Roane; TN Scott;
      TN Sevier.
054 Nashville, TN:
  SMSA Counties:
    1660 Clarksville - Hopkinsville, TN - KY_____18.2
      KY Christian; TN Montgomery.
    5360 Nashville - Davidson, TN_____15.8
      TN Cheatham, TN Davidson; TN Dickson; TN Robertson; TN Rutherford; TN Sumner;
      TN Williamson; TN Wilson.
    Non-SMSA Counties_____12.0
      KY Allen; KY Barren; KY Butler; KY Clinton; KY Cumberland; KY Edmonson;
      KY Logan; KY Metcalfe; KY Monroe; KY Simpson; KY Todd; KY Trigg; KY Warren;
      TN Bedford; TN Cannon; TN Clay; TN Coffee; TN DeKalb; TN Franklin; TN Giles;
      TN Hickman; TN Houston; TN Humphreys; TN Jackson; TN Lawrence; TN Lewis;
      TN Macon; TN Marshall; TN Maury; TN Moore; TN Overton; TN Perry; TN Pickett;

TN Putnam; TN Smith; TN Stewart; TN Trousdale; TN Van Buren; TN Warren; TN Wayne; TN White.

055 Memphis, TN:
  SMSA Counties:
    4920 Memphis, TN-AR-MS_____32.3
      AR Critteriden; MS Do Soto; TN Shelby; TN Tipton.
  Non-SMSA Counties_____26.5
    AR Clay; AR Craighead; AR Cross; AR Greene; AR Lawrence; AR Lee;
    AR Mississippi; AR Phillips; AR. Poinsett; AR Randolph; AR St. Francis; MS Alcorn;
    MS Benton; MS Bolivar; MS Calhoun; MS Carroll; MS Chickasaw, MS Clay;
    MS Coahoma; MS Grenada; MS Itawamba; MS Lafayette; MS Lee; MS Leflore;
    MS Marshall; MS Monroe; MS Montgomery; MS Panola; MS Pontotoc; MS Prentiss;
    MS Quitman; MS Sunflower; MS Tallahatchie; MS Tate; MS Tippah; MS Tishomingo;
    MS Union; MS Washington; MS Webster. MS Yalobusha; MO Dunklin;
    MO New Madrid; MO Pemiscot; TN Benton; TN Carroll; TN Chester; TN Crockett;
    TN Decatur; TN Dyer; TN Fayette; TN Gibson; TN Hardeman; TN Hardin;
    TN Haywood; TN Henderson; TN Henry; TN Lake; TN Lauderdale; TN McNairy;
    TN Madison; TN Obion; TN Weakley.

Kentucky:

056 Paducah, KY:
  Non-SMSA Counties_____5.2
    IL Hardin; IL Massac; IL Pope; KY Ballard; KY Caldwell; KY Calloway. KY Carlisle;
    KY Crittenden; KY Fulton; KY Graves; KY Hickman; KY Livingston; KY Lyon. KY
    McCracken; KY Marshall.
057 Louisville, KY:
  SMSA Counties:
    4520 Louisville, KY-IN_____11.2
      IN Clark; IN Floyd; KY Bullitt; KY Jefferson; KY Oldham.
  Non-SMSA Counties_____9.6
    IN Crawford; IN Harrison; IN Jefferson; IN Orange; IN Scott; IN Washington;
    KY Breckinridge; KY Grayson; KY Hardin; KY Hart; KY Henry; KY Larue; KY Marion;
    KY Meade; KY Nelson; KY Shelby; KY Spencer; KY Trimble; KY Washington.
058 Lexington, KY
  SMSA Counties
    4280 Lexington-Fayette, KY_____10.8
      KY Bourbon; KY Clark; KY Fayette; KY Jessamine; KY Scott; KY Woodford.
  Non-SMSA Counties_____7.0
    KY Adair KY Anderson; KY Bath; KY Boyle; KY Breathitt; KY Casey; KY Clay;
    KY Estill; KY Franklin; KY Garrard; KY Green; KY Harrison; KY Jackson; KY Knott;
    KY Lee; KY Leslie; KY Letcher; KY Lincoln; KY Madison; KY Magoffin; KY Menifee;
    KY Mercer; KY Montgomery; KY Morgan. KY Nicholas; KY Owsley; KY Perry;
    KY Powell; KY Pulaski; KY Rockcastle; KY Russell; KY Taylor; KY Wolfe.

West Virginia:

059 Huntington, WV:
  SMSA Counties:
    3400 Huntington – Ashland, WV-KY-OH_____2.9
      KY Boyd; KY Greenup; OH Lawrence; WV Cabell; WV Wayne.
  Non-SMSA Counties_____2.5
    KY Carter; KY Elliott; KY Floyd; KY Johnson; KY Lawrence; KY Martin; KY Pike;
    KY Rowan; OH Gallia; WV Lincoln; WV Logan; WV Mason; WV Mingo.
060 Charleston, WV
  SMSA Counties:

1480 Charleston, WV_____4.9
  WV Kanawha; WV Putnam.
Non-SMSA Counties_____4.2
  WV Boone; WV Braxton; WV Calhoun; WV Clay; Fayette; WV Gilmer; WV
  Greenbrier; WV Jackson; WV Monroe; WV Nicholas; WV Pocahontas; WV Raleigh;
  WV Roane; WV Summers; WV Webster; WV Wyoming.

061 Morgantown-Fairmont; WV:
Non-SMSA Counties_____2.1
  WV Barbour; WV Doddridge; WV Harrison; WV Lewis; WV Marion; WV Monongalia;
  WV Preston; WV Randolph; WV Taylor; WV Tucker, WV Upshur.

062 Parkersburg, WV:
SMSA Counties:
  6020 Parkersburg-Marietta. WV-OH_____1.1
    OH Washington; WV Wirt; WV Wood.
  Non-SMSA Counties_____1.2
    WV Pleasants; WV Ritchie.

063 Wheeling - Steubenville - Wierton, WV-OH:
SMSA Counties:
  8080 Steubenville-Wierton, OH-WV_____4.3
    OH Jefferson; WV Brooke; WV Hancock.
  9000 Wheeling, WV-OH_____2.4
    OH Belmont; WV Marshall; WV Ohio.
  Non-SMSA Counties_____3.0
    OH Harrison; OH Monroe; WV Tyler; WV Wetzel.

Ohio:

064 Youngstown-Warren, OH:
SMSA Counties:
  9320 Youngstown-Warren, OH_____9.4
    OH Mahoning; OH Trumbull.
  NonSMSA Counties_____6.7
    OH Columbiana; PA Lawrence; PA Mercer.

065 Cleveland, OH:
SMSA Counties:
  0080 Akron, OH_____7.8
    OH Portage; OH Summit.
  1320 Canton, OH_____6.1
    OH Carroll; OH Stark.
  1680 Cleveland, OH_____16.1
    OH Cuyahoga; OH Geauga; OH Lake; OH Medina.
  4440 Lorain-Elyria, OH_____9.3
    OH Lorain.
  4800 Mansfield, OH_____6.3
    OH Richland.
  Non-SMSA Counties:
    OH Ashland; OH Ashtabula; OH Coshocton; OH Crawford; OH Erie;
    OH Holmes; OH Huron; OH Tuscarawas; OH Wayne.

066 Columbus, OH:
SMSA Counties:
  1840 Columbus, OH_____10.6
    OH Delaware; OH Fairfield; Franklin; OH Madison; OH Pickaway.
  Non-SMSA Counties_____7.3
    OH Athens; OH Fayette; OH Guernsey; OH Hocking; OH Jackson; OH Knox;
    OH Licking; OH Marion; OH Meigs; OH Morgan; OH Morrow; OH Muskingum;
    OH Noble; OH Perry OH Pike; OH Ross; OH Scioto; OH Union; OH Vinton.

067 Cincinnati, OH:
SMSA Counties:
  1640 Cincinnati, OH-KY-IN_____11.0

IN Dearborn; KY Boone; KY Campbell; KY Kenton; OH Clermont; OH Hamilton; OH Warren.

3200 Hamilton-Middletown, OH_____5.0
   OH Butler.
Non-SMSA Counties_____9.2
   IN Franklin; IN Ohio; IN Ripley; IN Switzerland; KY Bracken; KY Carroll;
   KY Fleming; KY Gallatin; KY Grant; KY Lewis; KY Mason; KY Owen; KY Pendleton;
   KY Robertson; OH Adams; OH Brown; OH Clinton; OH Highland.

068 Dayton, OH:
  SMSA Counties:
   2000 Dayton, OH_____11.5
    OH Greene; ON Miami; OH Montgomery; OH Preble.
   7960 Springfield, OH_____7.8
    OH Champaign; OH Clark.
  Non-SMSA Counties_____9.9
   OH Darke; OH Logan; ON Shelby.

069 Lima, OH:
  SMSA Counties:
   4320 Lima, OH_____4.4
    OH Allen; OH Auglaize; OH Putnam; OH Van Wert.
  Non-SMSA Counties_____3.5
   OH Hardin; OH Mercer.

070 Toledo, OH:
  SMSA Counties:
   8400 Toledo, OH-MI_____8.8
    MI Monroe; OH Fulton; OH Lucas; OH Ottawa; OH Wood.
  Non-SMSA Counties_____7.3
   MI Lenawee; OH Hancock; OH Henry; OH Sandusky; OH Seneca; OH Wyandot.

Michigan:

071 Detroit, MI:,
  SMSA Counties:
   0440 Ann Arbor, MI_____8.5
    MI Washtenaw.
   2160 Detroit, MI_____17.7
    MI Lapeer; MI Livingston; MI Macomb; MI Oakland; MI St. Clair; Mi Wayne.
   2640 Flint, MI_____12.6
    MI Genesee; MI Shiawassee.
  Non-SMSA Counties_____16.7
   MI Sanilac.

072 Saginaw, MI:
  SMSA Counties:
   0800 Bay City, MI_____2.2
    MI Bay.
   6960 Saginaw, MI_____14.3
    MI Saginaw.
  Non-SMSA Counties_____5.2
   MI Alcona; MI Alpena; MI Arenac; MI Cheboygan; MI Chippewa; MI Clare;
   MI Crawford; MI Gladwin; MI Gratiot; MI Huron; MI Iosco; MI Isabella; MI Luce;
   MI Mackinac; MI Midland; MI Montmorency; MI Ogemaw; MI Oscoda; MI Otsego;
   MI Presque Isle; MI Roscommon; MI Tuscola.

073 Grand Rapids, MI:
  SMSA Counties:
   3000 Grand Rapids, MI_____5.2
    MI Kent; MI Ottawa.

5320 Muskegon - Norton Shores - Muskegon Heights, MI_____9.7
   MI Muskegon; MI Oceana.
Non-SMSA Counties_____4.9
   MI Allegan; MI Antrim; MI Benzie; MI Charlevoix; MI Emmet; MI Grand Traverse; MI
   Kalkaska; MI Lake; MI Leelanau; MI Manistee; MI Mason; MI Mecosta; MI Missaukee;
   MI Montcalm; MI Newaygo; MI Osceola; MI Wexford.
074 Lansing - Kalamazoo, MI:
  SMSA Counties:
   0780 Battle Creek, MI_____7.2
    MI Barry; MI Calhoun.
   3520 Jackson, MI_____5.1
    MI Jackson.
   3720 Kalamazoo-Portage, MI_____5.9
    MI Kalamazoo; MI Van Buren.
   4040 Lansing-East Lansing, MI_____5.5
    MI Clinton; MI Eaton; MI Ingham; MI Ionia.
  Non-SMSA Counties_____5.5
   MI Branch; MI Hillsdale.

Indiana:

075 South Bend, IN:
  SMSA Counties:
   7800 South Bend, IN_____7.1
    IN Marshall; IN St. Joseph,
   2330 Elkhart IN_____4.0
    IN Elkhart.
  Non-SMSA Counties_____6.2
   IN Fulton; IN Kosciusko; IN Lagrange; MI Berrien; MI Cass; MI St. Joseph.
076 Fort Wayne, IN:
  Non-SMSA Counties_____4.4
   IN Allen; IN Dekalb; IN Wells; IN Huntington; IN Noble; IN Steuben; IN Whitley;
   OH Defiance; OH Paulding; OH Williams.
077 Kokomo-Marion, IN:
  SMSA Counties:
   3850 Kokomo, IN_____4.4
    IN Howard; IN Tipton.
  Non-SMSA Counties_____3.7
   IN Cass; IN Grant; IN Miami; IN Wabash.
078 Anderson-Muncie, IN:
  SMSA Counties:
   0400 Anderson, IN_____4.9
    IN Madison.
   5280 Muncie, IN_____5.3
    IN Delaware.
  Non-SMSA Counties_____3.9
   IN Blackford; IN Fayette; IN Henry; IN Jay; IN Randolph; IN Union; IN Wayne.
079 Indianapolis, IN:
  SMSA Counties:
   1020 Bloomington, IN_____3.1
    IN Monroe.
   3480 Indianapolis, IN_____12.5
    IN Boone; IN Hamilton; IN Hendricks; IN Johnson; IN Marion; IN Morgan;
    IN Shelby.
  Non-SMSA Counties_____9.7
   IN Bartholomew; IN Brown; IN Daviess; IN Decatur; IN Greene; IN Jackson;
   IN Jennings; IN Lawrence; IN Martin; IN Owen; IN Putnam; IN Rush.

080 Evansville, IN:
  SMSA Counties
    2440 Evansville, IN-KY _____ 4.8
      IN Gibson; IN Posey; IN Vanderburgh; IN Warrick; KY Henderson.
    5990 Owensboro, KY _____ 4.7
      KY Daviess.
    Non-SMSA Counties _____ 3.5
      IL Edwards; IL Gallatin; IL Hamilton; IL Lawrence; IL Saline; IL Wabash;
      IL White; IN Dubois; IN Knox; IN Perry; IN Pike; IN Spencer; KY Hancock;
      KY Hopkins; KY McLean; KY Muhlenberg; KY Ohio; KY Union; KY Webster.
081 Terre Haute, IN:
  SMSA Counties:
    8320 Terre Haute, IN _____ 3.1
      IN Clay; IN Sullivan; IN Vermillion; IN Vigo.
    Non-SMSA Counties _____ 2.5
      IL Clark; IL Crawford; IN Parke.
082 Lafayette, IN:
  SMSA Counties:
    3920 Lafayette - West Lafayette, IN _____ 2.7
      IN Tippecanoe.
    Non-SMSA Counties _____ 1.5
      IN Benton; IN Carroll; IN Clinton; IN Fountain; IN Montgomery;
      IN Warren; IN White.

Illinois:

083 Chicago, IL:
  SMSA Counties:
    1600 Chicago, IL _____ 19.6
      IL Cook; IL Du Page; IL Kane; IL Lake; IL McHenry; IL Will.
    2960 Gary - Hammond - East Chicago, IN _____ 20.9
      IN Lake; IN Porter.
    3740 Kankakee. IL _____ 9.1
      IL Kankakee.
    3800 Kensoha, WI _____ 3.0
      WI Kenosha.
    Non-SMSA Counties _____ 18.4
      IL Bureau; IL De Kalb; IL Grundy; IL Iroquois; IL Kendall; IL La Salle;
      IL Livingston; IL Putnam; IL Jasper; IN Laporte; IN Newton; IN Pulaski; IN Starke.
084 Champaign-Urbana, IL:
  SMSA Counties:
    1400 Champaign - Urbana – Rantoul, IL _____ 7.8
      IL Champaign.
    Non-SMSA Counties _____ 4.8
      IL Coles; IL Cumberland; IL Douglas; IL Edgar; IL Ford; IL Platt; IL Vermilion.
085 Springfield-Decatur, IL:
  SMSA Counties:
    2040 Decatur, IL _____ 7.6
      IL Macon.
    7880 Springfield, IL _____ 4.5
      IL Menard; IL Sangamon.
    Non-SMSA Counties _____ 4.0
      IL Cass; IL Christian; IL De Witt; IL Logan; IL Morgan; IL Moultrie;
      IL Scott; IL Shelby.
086 Quincy, IL:
  Non-SMSA Counties _____ 3.1
    IL Adams; IL Brown; IL Pike; MO Lewis; MO Marlon; MO Pike; MO Rails.

---

087 Peoria, IL:
  SMSA Counties
    1040 Bloomington - Normal, IL_____2.5
      IL McLean.
    8120 Peoria, IL_____4.4
      IL Peoria; IL Tazewell; IL Woodford.
    Non-SMSA Counties_____3.3
      IL Fulton; IL Knox; IL McDonough; IL Marshall; IL Mason; IL Schuyler;
      IL Stark; IL Warren.
088 Rockford, IL:
  SMSA Counties:
    6880 Rockford, IL_____6.3
      IL Boone; IL Winnegago.
    3620 Janesville - Beloit WI_____3.1
      WI Rock
    Non-SMSA Counties_____4.6
      IL Lee; IL Ogle; IL Stephenson.

Wisconsin:

089 Milwaukee, WI:
  SMSA Counties:
    5080 Milwaukee, WI_____8.0
      WI Milwaukee; WI Ozaukee; WI Washington; WI Waukesha.
    6600 Racine, WI_____8.4
      WI Racine.
    Non-SMSA Counties_____7.0
      WI Dodge; WI Jefferson; WI Sheboygan; WI Walworth.
090 Madison, WI:
  SMSA Counties:
    4720 Madison, WI_____2.2
      WI Dane.
    Non-SMSA Counties_____1.7
      WI Adams; WI Columbia; WI Green; WI Iowa; WI Marquette; WI Richland; WI Sauk.
091 La Crosse, WI:
  SMSA Counties:
    3870 LaCrosse. WI_____0.9
    Non-SMSA Counties_____0.6
      MN Houston; MN Winona; WI Buffalo; WI Jackson; WI Juneau; WI Monroe;
      WI Trempealeau; WI Vernon.
092 Eau Claire, WI:
  SMSA Counties:
    2290 Eau Claire, WI_____0.5
      WI Chippewa; WI Eau Claire.
    Non-SMSA Counties_____0.6
      WI Barron; WI Dunn; WI Pepin; WI Rusk; WI Sawyer; WI Washburn.
093 Wausau, WI:
  Non-SMSA Counties_____0.6
      WI Clark; WI Langlade; WI Lincoln; WI Marathon; WI Oneida; WI Portage;
      WI Price; WI Taylor; WI Vilas; WI Wood.
094 Appleton - Green Bay - Oshkosh, WI:
  SMSA Counties:
    0460 Appleton-Oshkosh, WI_____0.9
      WI Calumet; WI Outaramie; WI Winnebago.
    3080 Green Bay, WI_____1.3
      WI Brown.

Non-SMSA Counties _____ 1.0
  MI Alger; MI Baraga; MI Delta; MI Dickinson; MI Houghton; MI Iron;
  MI Keweenaw; MI Marquette; MI Menominee; MI Schoolcraft; WI Door;
  WI Florence; WI Fond Du Lac; WI Forest WI Green Lake; WI Kewaunea;
  WI Manitowoc; WI Marinette; WI Menominee; WI Oconto; WI Shawano;
  WI Waupaca; Waushara.

095 Duluth, MN:
  SMSA Counties:
    2240 Duluth - Superior, MN-WI _____ 1.0
      MN St Louis; WI Douglas.
  Non-SMSA Counties _____ 1.2
    MI Gogebic; MI Ontonagon; MN Carlton; MN Cook; MN Itasca; MN Koochiching;
    MN Lake; WI Ashland; W! Bayfield; WI Iron.

Minnesota:

096 Minneapolis-St. Paul, MN:
  SMSA Counties:
    5120 Minneapolis-St. Paul, MN-WI _____ 2.9
      MN Anoka; MN Carver; MN Chisago; MN Dakota; MN Hennepin; MN Ramsey;
      MN Scott; MN Washington; MN Wright; MN St. Croix.
    6980 St. Cloud, MN _____ 0.5
      MN Benton; MN Sherburne; MN Stearns.
  Non-SMSA Counties _____ 2.2
    MN Aitkin; MN Big Stone; MN Blue Earth; MN Brown; MN Cass; MN Chippewa;
    MN Crow Wing; MN Douglas; MN Faribault; MN Goodhue; MN Grant; MN Isanti;
    MN Kanabec; MN Kandiyohi; MN Lac Qui Parle; MN Le Sueur; MN McLeod;
    MN Martin; MN Meeker; MN Mille Lacs; MN Mornson; MN Nicollet; MN Pine;
    MN Pope; MN Renville; MN Rice; MN Sibley; MN Stevens; MN Swift; MN Todd;
    MN Traverse; MN Wadena; MN Waseca; MN Watonwan; MN Yellow Medicine;
    WI Burnett; WI Pierce; WI Polk.

097 Rochester, MN:
  SMSA Counties:
    6820 Rochester, MN _____ 1.4
      MN Olmsted.
  Non-SMSA Counties _____ 0.9
    MN Dodge; MN Fillmore; MN Freeborn; MN Mower; MN Steele; MN Wabasha.

Iowa:

098 Dubuque, IA:
  SMSA Counties:
    2200 Dubuque, IA _____ 0.6
      IA Dubuque
  Non-SMSA Counties _____ 0.5
    IL Jo Daviess; IA Allamakee; IA Clayton; IA Delaware, IA Jackson;
    IA Winneshiek; WI Crawford; WI Grant; WI Lafayette.

099 Davenport-Rock Island-Moline, IA-IL:
  SMSA Counties:
    1960 Davenport-Rock Island-Moline, IA-IL _____ 4.6
      IL Henry; IL Rock Island; IA Scott.
  Non-SMSA Counties _____ 3.4
    IL Carroll; IL Handcock; IL Henderson; IL Mercer; IL Whiteside; IA Clinton;
    IA Des Moines; IA Henry; IA Lee; IA Louisa; IA Muscatine; MO Clark.

100 Cedar Rapids, IA:
  SMSA Counties:
    1360 Cedar Rapids, IA _____ 1.7
      IA Linn.
  Non-SMSA Counties _____ 1.5

IA Benton; IA Cedar; IA Iowa; IA Johnson; IA Jones; IA Washington.

101 Waterloo, IA:
  SMSA Counties:
    8920 Waterloo-Cedar Falls, IA_____4.7
    IA Black Hawk.
  Non-SMSA Counties_____2.0
    IA Bremer; IA Buchanan; IA Butler; IA Cerro Gordo; IA Chickasaw; IA Fayette;
    IA Floyd; IA Franklin; IA Grundy; IA Hancock; IA Hardin; IA Howard; IA Mitchell;
    IA Winnegago; IA Worth.

102 Fort Dodge, IA:
  Non-SMSA Counties_____0.4
    IA Bueno Vista; IA Calhoun; IA Carroll; IA Clay; IA Dickinson; IA Emmet;
    IA Greene; IA Hamilton; IA Humboldt; IA Kossuth; IA Palo Alto; IA Pocahontas;
    IA Sac; IA Webster; IA Wright.

103 Sioux City, IA:
  SMSA Counties:
    7720 Sioux City, IA-NE_____1.9
    IA Woodbury; NE Dakota.
  Non-SMSA Counties_____1.2
    IA Cherokee, IA Crawford; IA Ida; IA Monona; IA O'Brien; IA Plymouth; IA. Sioux;
    NE Antelope; NE Cedar; NE Cuming; NE Dixon; NE Knox; NE Madison; NE Pierce;
    NE Stanton; NE Thurston; NE Wayne; SD Bon Homme; SD Clay; SD Union;
    SD Yankton.

104 Des Moines, IA:
  SMSA Counties:
    2120 Des Moines, IA_____4.5
    IA Polk; IA Warren.
  Non-SMSA Counties_____2.4
    IA Adair; IA Appanoose; IA Boone; IA Clarke; IA Dallas; IA Davis; IA Decatur;
    IA Guthrie; IA Jasper; IA Jefferson; IA Keokuk; IA Lucas; IA Madison; IA Mahaska;
    IA Marion; IA Marshall; IA Monroe; IA Poweshiek; IA Ringgold; IA Story; IA Tama;
    IA Union; IA Van Buren; IA Wapello; IA Wayne.

Missouri:

105 Kansas City, MO:
  SMSA Counties:
    3760 Kansas City, MO-KS_____12.7
    KS Johnson; KS Wayandotte; MO Cass; MO Clay; MO Jackson; MO Platte; MO Ray.
    4150 Lawrence, KS_____7.2
    7000 St Joseph. MO_____3.2
    MO Andrew; MO Buchanan.
  Non-SMSA Counties_____10.0
    KS Anderson; KS Atchison; KS Brown; KS Doniphan; KS Franklin; KS Leavenworth;
    KS Linn; KS Miami; MO Atchison; MO Bates; MO Benton; MO Caldwell; MO Caroll;
    MO Clinton; MO Daviess; MO Dekalb; MO Gentry; MO Grundy; MO Harrison;
    MO Henry; MO Holt; MO Johnson; MO Lafayette; MO Livingston; MO Mercer;
    MO Nodaway; MO Pettis; MO Saline; MO Worth.

106 Columbia, MO:
  SMSA Counties:
    1740 Columbia, MO; MO Boone _____6.3
  Non-SMSA Counties_____4.0
    MO Adrain; MO Audrain; MO Callaway; MO Camden; MO Chariton; MO Cole;
    MO Cooper; MO Howard; MO Knox; MO Linn;. MO Macon; MO Miller; MO Moniteau;
    MO Monroe; MO Morgan; MO Osage; MO Putnam; MO Randolph; MO Schuyler; MO
    Scotland; MO Shelby; MO Sullivan.

107 St. Louis, MO:
    SMSA Counties:
        7040 St. Louis, MO-IL_____14.7
            IL Clinton; IL Madison; IL Monroe; IL St. Clair; MO Franklin; MO Jefferson; MO St.
            Charles; MO St. Louis; MO St. Louis City.
    Non-SMSA Counties_____11.4
        IL Alexander IL Bond; IL Calhoun; IL Clay; IL Effingharn; IL Fayette; IL Franklin;
        IL Greene; IL Jackson; IL Jasper; IL Jefferson; IL Jersey; IL Johnson; IL Macoupin;
        IL Marion; IL Montgomery; IL Perry; IL Pulaski; IL Randolph; IL Richland; IL Union;
        IL Washington; IL Wayne; IL Williamson; MO Bollinger; MO Butler;
        MO Cape Girardeau; MO Carter; MO Crawford; MO Dent; MO Gasconade; MO Iron;
        MO Lincoln; MO Madison; MO Maries; Mississippi; MO Montgomery; MO Perry;
        MO Phelps; MO Reynolds; MO Ripley; MO St. Francis; MO Ste. Genevieve; MO Scott;
        MO Stoddard; MO Warren; MO Washington; MO Wayne.
108 Springfield, MO:
    SMSA Counties:
        7920 Springfield, MO_____2.0
            MO Christian; MO Greene.
    Non-SMSA Counties_____2.3
        KS Allen; KS Bourbon; KS Cherokee; KS Crawford; KS Labette; KS Montgomery; KS
        Neosho; KS Wilson; KS Woodson; MO Barry; MO Barton; MO Cedar; MO Dade; MO
        Dallas;.MO Douglas; MO Hickory; MO Howell; MO Jasper; MO Laclede; MO Lawrence;
        MO McDonald; MO Newton; MO Oregon; MO Ozark; MO Polk; MO Pulaski;
        MO St. Clair; MO Shannon; MO Stone; MO Taney; MO Texas; MO Vernon;
        MO Webster; MO Wright; OK Craig; OK Ottawa.

Arkansas:

109 Fayetteville, AR:
    Non-SMSA Counties_____3.3
        AR Baxter; AR Benton; AR Boone; AR Carroll; AR Madison; AR
        Marion; AR Newton; AR Searcy; AR Washington; OK Adair; OK
        Delaware.
110 Fort Smith, AR:
    SMSA Counties:
        2720 Fort Smith, AR-OK_____5.6
            AR Crawford; AR Sebastian; OK Le Flore; OK Sequoyah.
    Non-SMSA Counties_____6.6
        AR Franklin; AR Logan; AR Polk; AR Scott; OK Choctaw; OK Haskell; OK Latimer; OK
        McCurtain; OK Pittsburg; OK Pushmataha.
111 Little Rock-North Little Rock, AR:
    SMSA Counties:
        4400 Little Rock-North Little Rock, AR_____15.7
            AR Pulaski; AR Saline.
        6240 Pine Bluff, AR_____31.2
            AR Jefferson
    Non-SMSA Counties_____16.4
        AR Arkansas; AR Ashley; AR Bradley; AR Calhoun; AR Chicott; AR Clark; AR
        Calhoun; AR Cleveland; AR Conway; AR Dallas; AR Desha; AR Drew; AR Faulkner;
        AR Fulton: AR Garland; AR Grant; AR Hot Springs; AR Independence; AR Izard; AR
        Jackson; AR Johnson; AR Lincoln; AR Lonoke; AR Monroe; AR Montgomery; AR
        Ouachita; AR Perry, AR Pope; AR Prairie; AR Sharp; AR Stone; AR Union; AR Van
        Buren; AR While; AR Woodruft; AR Yell.

Mississippi:

112 Jackson, MS:
  SMSA Counties;
    3560 Jackson, MS_____30.3
    MS Hinds; MS Rankin.
  Non-SMSA Counties_____32.0
    MS Attala; MS Choctaw; MS Choctaw; MS Clarke; MS Copiah; MS Covington; MS
    Franklin; MS Holmes: MS Humphreys; MS Issaquena; MS Jasper; MS Jefferson; MS
    Jefferson Davis; MS Jones; MS Kemper; MS Lauderdale; MS Lawrence; MS Leake;
    MS Lincoln; MS Lowndes; MS Madison; MS Neshoba; MS Newton; MS Noxubee; MS
    Oktibbeha; MS Scott; MS Sharkey; MS Simpson; MS Smith; MS Warren; MS Wayne;
    MS Winston; MS Yazoo.

Louisiana:

113 New Orleans, LA:
  SMSA Counties
    0920 Biloxi-Gulfport, MS_____19.2
    MS Hancock; MS Harrison; MS Stone.
    5560 New Orleans, LA_____31.0
    LA Jefferson; LA Orleans; LA St. Bernard; LA St. Tammany.
  Non-SMSA Counties_____27.7
    LA Assumption; LA Lafourche; LA Plaquemines; LA St. Charles; LA St. James;
    LA St. John The Baptist; LA Tangipahoa; LA Terrebonne; LA Washington; MS Forrest;
    MS Lamar; MS Marion; MS Pearl River; MS Perry; MS Pike; MS Walthall.
114 Baton Rouge, LA:
  SMSA Counties:
    0760 Baton Rouge, LA_____26.1
    LA Ascension; LA East Baton Rouge; LA Livingston; LA West Baton Rouge.
  Non-SMSA Counties_____30.4
    LA Concordia; LA E. Feliciana; LA Iberville; LA Pointe Coupee; LA St. Helena;
    LA West Feliciana; MS Adams; MS Amite; MS Wilkinson.
115 Lafayette, LA:
  SMSA Counties:
    3880 Lafayette, LA_____20.6
    LA Lafayette.
  Non-SMSA Counties_____24.1.
    LA Acadia; LA Evangeline; LA Iberia; LA St. Landry; LA St. Martin;
    LA St. Mary; LA Vermillion.
116 Lake Charles, LA:
  SMSA Counties:
    3960 Lake Charles, LA_____19.3
    LA Calcasieu.
  Non-SMSA Counties_____17.8
    LA Allen; LA Beauregard; LA Cameron; LA Jefferson Davis LA Vernon.
117 Shreveport, LA:
  SMSA Counties:
    0220 Alexandria, LA_____25.7
    LA Grant; LA Rapides.
    7680 Shreveport, LA_____29.3
    LA Bossier; LA Caddo; LA Webster.
  Non-SMSA Counties_____29.3
    LA Avoyelles; LA Bienville; LA Claiborne; LA De Soto; LA Natchitoches;
    LA Red River; LA Sabine; LA Winn.

118 Monroe, LA:
  SMSA Counties:
    5200 Monroe, LA _____22.8
      LA Ouachita.
    Non-SMSA Counties _____27.9
      LA Caldwell; LA Catahoula; LA East Carroll; LA Franklin; LA Jackson; LA La Salle; LA
      Lincoln; LA Madison; LA Morehouse; LA Richland; LA Tensas; LA Union; LA West
      Carroll.

Texas:

119 Texarkana, TX:
  SMSA Counties:
    8360 Texarkana, TX-Texarkana, AR _____19.7
      AR Little River; AR Miller; TX Bowie.
    Non-SMSA Counties _____20.2
      AR Columbia; AR Hempstead; AR Howard; AR Lafayette; AR Nevada; AR Pike; AR
      Sevier; TX Camp; TX Cass; TX Lamar; TX Morris; TX Red River; TX Titus.
120 Tyler-Longview, TX:
  SMSA Counties:
    4420 Longview, TX _____22.8
      TX Gregg; TX Harrison.
    8640 Tyler, TX _____23.5
      TX Smith.
    Non-SMSA Counties _____22.5
      TX Anderson; TX Angelina; TX Cherokee; TX Henderson; TX Houston; TX Marion; TX
      Nacogdoches; TX Panola; TX Rusk; TX San Augustine; TX Shelby; TX Upshur; TX
      Wood.
121 Beaumont-Port Arthur, TX:
  SMSA Counties:
    0840 Beaumont-Port Arthur Orange, TX _____22.6
      TX Hardin; TX Jefferson; TX Orange.
    Non-SMSA Counties _____22.6
      TX Jasper; TX Newton; TX Sabine; TX Tyler.
122 Houston, TX:
  SMSA Counties
    1260 Bryan-College Station, TX _____23.7
      TX Brazos.
    2920 Galveston-Texas City, TX _____28.9
      TX Galveston.
  3360 Houston, TX _____27.3
      TX Brazona; TX Fort Bend; TX Harris; TX Liberty, TX Montgomery, TX Waller.
    Non-SMSA Counties _____27.4
      TX Austin; TX Burleson; TX Calhoun; TX Chambers; TX Colorado; TX De Witt; TX
      Fayette; TX Goliad; TX Grimes; TX Jackson; TX Lavaca; TX Leon; TX Madison; TX
      Matagorda; TX Polk; TX Robertson; TX San Jacinto; TX Trinity; TX Victoria; TX Walker;
      TX Washington; TX Wharton.
123 Austin, TX:
  SMSA Counties:
  0640 Austin, TX _____24.1
      TX Hays; TX Travis; TX Williamson.
    Non-SMSA Counties _____24.2
      TX Bastrop; TX Blanco; TX Burnet; TX Caldwell; TX Lee; TX Llano.
124 Waco-Killeen-Temple, TX:
  SMSA Counties:
  3810 Killeen-Temple, TX. _____16.4
      TX Belt TX Coryall.
  8800 Waco, TX _____20.7
      TX McLermarx

Non-SMSA Counties _____ 18.6
   TX Bosque; TX Falls; TX Freestone; TX Hamilton; TX Hill; TX Lampasas; TX
   Limestone; TX Milam; TX Mills.

125 Dallas-Fort Worth, TX:
  SMSA Counties
   1920 Dallas-Fort Worth, TX _____ 18.2
    TX Collier; TX Dallas; TX Denton; TX Ellis; TX Hood; TX Johnson; TX
    Kaufman; TX Parker; TX Rockwall; TX Tarrant; TX Wise.
   7640 Sherman-Denison, TX _____ 9.4
    TX Grayson.
  Non-SMSA Counties _____ 17.2
   OK Bryan; TX Cooke; TX Delta; TX Erath; TX   Fannin; TX Franklin; TX Hopkins; TX
   Hunt; TX Jack; TX Montague; TX Navarro; TX Palo Pinto; TX Rains; TX Sommerveil; TX
   Van Zandt.

126 Wichita Falls, TX:
  SMSA Counties:
   9080 Wichita Falls, TX: _____ 12.4
    TX Clay; TX Wichita.
  Non-SMSA Counties _____ 11.0
   TX Archer; TX Baylor; TX Cottle; TX Foard; TX Hardeman; TX Wilbarger; TX Young.

127 Abilene, TX:
  SMSA Counties:
   0040 Abilene, TX _____ 11.6
    TX Callahan; TX Jones; TX Taylor.
  Non-SMSA Counties _____ 10.9
   TX Brown; TX Coleman; TX; Comanche; TX Eastland; TX Fisher; TX Haskell; TX Kent;
   TX Knox; TX Mitchell; TX Nolan; TX Scurry; TX Shackelford; TX Stephens; TX
   Stonewall; TX Throckmorton.

128 San Angelo, TX:
  SMSA Counties:
   7200 San Angelo, TX _____ 19.2
    TX Tom Green.
  Non-SMSA Counties _____ 20.0
   TX Coke; TX Concha; TX Crockett; TX Irion; TX Kimble; TX McCulloch; TX Mason; TX
   Menard; TX Reagan; TX Runnels; TX San Saba; TX Schleicher; TX Sterling; TX
   Sutton, TX Terrell.

129 San Antonio, TX:
  SMSA Counties:
   4080 Laredo _____ 87.3
    TX Webb.
   7240 San Antonio, TX _____ 47.8
    TX Bexar; TX Comal; TX Guadalupe.
  Non-SMSA Counties _____ 49.4
   TX Atascosa; TX Bandera; TX Dimmit; TX Edwards; TX Frio; TX Gillespie; TX
   Gonzales; TX Jim Hogg; TX Karnes; TX Kendall; TX Kerr; TX Kinney; TX La Salle; TX
   McMullen; TX Maverick; TX Medina; TX Real; TX Uvalde; TX Val Verde; TX Wilson; TX
   Zapata; TX Zavala.

130 Corpus Christi, TX:
  SMSA Counties:
   1880 Corpus Christi, TX _____ 41.7
    TX Nueces; TX San Patricio.
  Non-SMSA Counties _____ 44.2
   TX Aransas; TX Bee; TX Brooks; TX Duval; TX Jim Wells; TX Kenady; TX Kyberg; TX
   Live Oak; TX Refugio.

131 Brownsville-McAllen-Harlingen, TX:
  SMSA Counties:
   1240 Brownsville-Harlingen-San Benito, TX _____ 71.0
    TX Cameron.
   4880 McAllen-Pharr-Edinburg, TX _____ 72.8

TX Hidalgo.

Non-SMSA Counties _____72.9
  TX Starr; TX Willacy.

132 Odessa-Midland, TX:
  SMSA Counties:
    5040 Midland, TX _____19.1
      TX Midland.
    5800 Odessa, TX _____15.1
      TX Ector.
  Non-SMSA Counties _____18.9
    TX Andrews; TX Crane; TX Glasscock; TX Howard; TX Loving; TX Martin; TX Pecos;
    TX Reeves; TX Upton; TX Ward; TX Winkler.

133 El Paso, TX:
  SMSA Counties:
    2320 El Paso, TX _____57.8
      TX El Paso.
  Non-SMSA Counties _____49.0
    NM Chaves; NM Dona Ana; NM Eddy; NM Grant; NM Hidalgo; NM Luna; NM Otero;
    NM Sierra, TX Brewster; TX Culberson; TX Hudspeth; TX Jeff Davis; TX Presidio.

134 Lubbock, TX:
  SMSA Counties:
    4600 Lubbock_____19.6
      TX Lubbock.
  Non-SMSA_____19.5
    NM Lea; NM Roosevelt ; TX Bailey; TX Borden; TX Cochran; TX Crosby; TX Dawson;
    TX Dickens; TX Floyd; TX Gaines; TX Garza; TX Hale; TX Hockley; TX King; TX Lamb;
    TX Lynn; TX Motley; TX Terry; TX Yoakum.

135 Amarillo, TX:
  SMSA Counties:
    0320 Amarillo, TX _____9.3
      TX Potter; TX Randall.
  Non-SMSA Counties _____11.0
    NM Curry; NM Harding; NM Quay; NM Union; OK Beaver; OK Cimarron; OK Texas; TX
    Arnstrong; TX Briscoe; TX Carson; TX Castro; TX Childress; TX Collingsworth; TX
    Dallam; TX Deaf Srnith; TX Donley; TX Gray; TX Hall; TX Hansford; TX Hartley; TX
    Hemphill; TX Hutchinson; TX Lipscomb; TX Moore; TX Ochitree; TX Oldham; TX
    Parmer; TX Roberts; TX Sherman; TX Swisher; TX Wheeler.

Oklahoma:

136 Lawton, OK:
  SMSA Counties:
    4200 Lawton, OK _____14.8
      OK Comanche.
  Non-SMSA Counties _____10.8
    OK Cotton; OK Green; OK. Harmon; OK Jackson; OK Jefferson; OK Kiowa; OK
    Stephens; OK Tillman.

137 Oklahoma City, OK:
  SMSA Counties
    5880 Oklahoma City, OK _____10.2
      OK Canadian; OK Cleveland; OK McClain; OK Oklahoma; OK Pottawatomie.
  Non-SMSA Counties_____9.0
    OK Alfalfa; OK Atoka; OK Beckham; OK Blaine; OK Caddo; OK Carter; OK Coat; OK
    Custer; OK Dewey; OK Ellis; OK Garfield; OK Garvin; OK Grady; OK Grant; OK
    Harper; OK Hughes; OK Johnston; OK Kingfisher; OK Lincoln; OK Logan; OK Love;
    OK Major; OK Marshall; OK Murray, OK Okfuskee; OK Pontotoc; OK Roger Mills; OK
    Seminole; OK Washita; OK Woods; Ok Woodward.

138 Tulsa, OK:
  SMSA Counties:

8560 Tulsa, OK _____ 10.2
    OK Creek; OK Mayes; OK Osage; OK Rogers; OK Tulsa; OK Wagoner.
Non-SMSA Counties _____ 10.0
    OK Cherokee; OK Key; OK McIntosh; OK Muskogee; OK Noble; OK Nowata; OK
    Okmulgee; OK Pawnee; OK Payne; OK Washington.

Kansas:

139 Wichita, KS:
  SMSA Counties:
    9040 Wichita, KS_____ 7.9
      KS Butler; KS Sedgwick.
  Non-SMSA Counties _____ 5.7
    KS Barber; KS Barton; KS Chase; KS Chautauqua; KS Clark; KS Comanche. KS
    Cowley; KS Edwards; KS Elk; KS Finney; KS Ford; KS Grant; KS Gray; KS Greeley;
    KS Greenwood; KS Hamilton; KS Harper; KS Harvey; KS Haskell; KS Hodgeman; KS
    Kearny; KS Kingman; KS Kiowa; KS Lane; KS McPherson; KS Marion; KS Meade; KS
    Morton; KS Ness; KS Pawnee; KS Pratt; KS Reno; KS Rice; KS Rush; KS Scott; KS
    Seward; KS Stafford; KS Stanton; KS Stevens; KS Sumner, KS Wichita.
140 Salina, KS:
  Non-SMSA Counties _____ 1.5
    KS Cheyenne; KS Cloud; KS Decatur; KS Dickinson; KS Ellis; KS Ellsworth; KS Gove;
    KS Graham; KS Jewell; KS Lincoln; KS Logan; KS Mitchell; KS Norton; KS Osborne;
    KS Ottawa; KS Phillips; KS Rawlins; KS Republic; KS Rooks; KS Russell; KS Saline;
    KS Sheridan; KS Sherman; KS Smith; KS Thomas; KS Trego; KS Wallace.
141 Topeka, KS:
  SMSA Counties:
    8440 Topeka, KS _____ 9.0
      KS Jefferson; KS Osage; KS Shawnee.
  Non-SMSA Counties _____ 6.5
    KS Clay; Coffey; KS Geary; KS Jackson; KS Lyon; KS Marshall; KS Morris; KS
    Nemaha; KS Pottawatomie, KS Riley; KS Wabaunsee; KS Washington.

Nebraska:

142 Lincoln, NE:
  SMSA Counties:
    4360 Lincoln, NE _____ 2.8
      NE Lancaster.
  Non SMSA Counties _____ 1.9
    NE Butler; NE Fillmore; NE Gage; NE Jefferson; NE Johnson; NE Nemaha; NE Otoe;
    NE Pawnee; NE Polk; NE Richardson; NE Saline, NE Seward; NE Thayer; NE York.
143 Omaha, NE:
  SMSA Counties:
    5920 Omaha, NE-IA _____ 7.6
      IA Pottawattamie; NE Douglas; NE Sarpy.
  Non-SMSA _____ 5.3
    IA Adams; IA Audubon; IA Cass; IA Fremont; IA Harrison; LA Mills; IA Montgomery; IA
    Page; IA Shelby; IA Taylor; NE Burt; NE Cass; NE Colfax; NE Dodge; NE Platte; NE
    Saunders; NE Washington.
144 Grand Island, NE:
  Non SMSA Counties_____ 1.4
    NE Adams; NE Aurther; NE Blaine; NE Boyd; NE Brown; NE Buffalo; NE Chase; NE Cherry;
    NE Clay; NE Custer; NE Dawson; NE Dundy; NE Franklin; NE Frontier; NE Fumas; NE
    Garfield; NE Gosper; NE Grant; NE Greeley, NE Hall; NE Hamilton; NE Harlan; NE Hayes; NE
    Hitchcock; NE Holt; NE Hooker; NE Howard; NE Kearney; NE Keith; NE Keya Paha; NE
    Lincoln; NE Logan; NE Loup; NE McPherson; NE Merrick; NE Nance; NE Nuckolls; NE
    Perkins; NE Phelps; NE Red Willow; NE Rock; NE Sherman; NE Thomas; NE Valley; NE
    Webster; NE Wheeler.
145 Scottsbluff, NE:

Non-SMSA Counties _____ 5.3
    NE Banner; NE Box Butt; NE Cheyenne; NE Dawes; NE Deuel; NE Garden; NE Kimball; NE Morrill; NE Scotts Buff; NE Sheridan; NE Sioux; NE Goshen.

South Dakota:

146 Rapid City, SD:
  SMSA Counties:
    6660 Rapid City, SD_____ 3.4
      SD Pennington; SD Meade.
  Non-SMSA Counties_____ 7.9
    SD Bennett; SD Buffalo; SD Butte; SD Campbell; SD Corson; SD Custer; SD Dewey (Armstrong); SD Fall River; SD Haakon; SD Harding; SD Hughes; SD Hyde; SD Jackson; SD Jones; SD Lawrence; SD Lyman; SD Mellette; SD Perkins; SD Potter; SD Shannon (Washington); SD Stanley; SD Sully; SD Todd; SD Tripp; SD Walworth; SD Washabaugh; SD Ziebach; WY Crook; WY Niobrara; WY Weston.
147 Sioux Falls, SD:
  SMSA Counties:
    7760 Sioux Falls, SD_____ 1.2
      SD Minnehaha.
  Non-SMSA Counties_____ 0.8
    IA Lyon; IA Osceola; MN Cottonwood; MN Jackson;. MN Lincoln; MN Lyon; MN Murray, MN Nobles; MN Pipestone; MN Redwood; MN Rock; SD Aurora; SD Beadle; SD Brookings; SD Brule; SD Charles Mix; SD Davison; SD Douglas; SD Gregory; SD Hand; SD Hanson; SD Hutchinson; SD Jerauld; SD Kingsbury; SD Lake; SD Lincoln; SD McCook, SD Miner, SD Moody, SD Sanborn; SD Turner.
148 Aberdeen, SD:
  Non-SMSA Counties_____ 1.3
    SD Brown; SD Clark; SD Codington; SD Day; SD Deuel; SD Edmunds; SD Faulk; SD Grant; SD Hamlin; SD McPherson; SD Marshall; SD Roberts; SD Spink.

North Dakota:

149 Fargo-Moorhead, ND-MN:
  Non-SMSA Counties_____ 0.7
    MN Becker MN Clay; MN Cass; MN Wilkin; ND Barnes; ND Dickey; ND Eddy; ND Foster; ND Griggs; ND La Moure; ND Logan; ND McIntosh; ND Ransom; ND Richland; ND Sargent; ND Steele; ND Stutsman; ND Traill.
150 Grand Forks, ND:
  SMSA Counties:
    2985 Grand Forks, ND-MN_____ 1.2
      MN Polk; ND Grand Forks.
  Non-SMSA Counties_____ 2.0
    MN Beltrami; MN Clearwater MN Hubbard. MN Kittson; MN Lake of the Woods; MN Mahnomen; MN Marshall; MN Norman; MN Pennington; MN Red Lake; MN Roseau; MN Benson; ND Cavalier; ND Nelson; ND Pembina; ND Ramsey; ND Towner; ND Walsh.
151 Bismarck, ND:
  SMSA Counties:
    1010 Bismarck, ND_____ 0.4
    ND Burleigh; ND Morton.

Non-SMSA Counties_____1.3
 ND Adams; ND Billings; ND Bowman; ND Dunn; ND Emmons; ND Golden Valley;
 ND Grant; ND Hettinger; ND Kidder; ND Mercer; ND Oliver; ND Sheridan; ND Sioux;
 ND Slope; ND Stark; ND Wells.
152 Minot, ND:
 Non-SMSA Counties_____4.4
 MT Daniels; MT Richland; MT Roosevelt; MT Sheridan; ND Bottineau; ND Burke;
 ND Divide; ND McHenry; ND McKenzie; ND McLean; ND Mountrail; ND Pierce;
 ND Renville; ND Rolette; ND Ward; ND Williams.

Montana:

153 Great Falls, MT:
 SMSA Counties.
  3040 Great Falls, MT_____3.2
  MT Cascade.
 Non-SMSA Counties_____4.1
  MT Blaine; MT Broadwater; MT Chouteau; MT Fergus; MT Glacier; MT Hill;
  MT Jefferson; MT Judith Basin; MT Lewis and Clark; MT Liberty; MT Meagher;
  MT Petroleum; MT Phillips; MT Pondera; MT Teton; MT Toole; MT Valley;
  MT Wheatland.
154 Missoula, MT:
 Non-SMSA Counties_____2.7
  MT Beaverhead; MT Deer Lodge; MT Flathead; MT Granite; MT Lincoln;
  MT Madison; MT Mineral; MT Missoula; MT Powell; MT Ravalli; MT Sanders;
  MT Silver Bow; MT Lake.
155 Billings, MT:
 SMSA Counties:
  0880 Billings, MT_____3.3
  MT Yellowstone.
 Non-SMSA Counties_____3.3
  MT Big Horn; MT Carbon; MT Carter; MT Custer; MT Dawson; MT Fallon; MT
  Gallatin; MT Garfield; MT Golden Valley; MT McCone; MT Musselshell; MT Park; MT
  Powder River; MT Prairie; UT Rosebud; MT Stillwater, MT Sweet Grass; MT
  Treasure; MT Wilbaux; MT Yellowstone Nat'l Park; WY Big Horn; WY Hot Springs;
  WY Park; WY Sheridan; WY Washakie.

Wyoming:

156 Cheyenne-Casper, WY:
 Non-SMSA Counties_____7.5
  CO Jackson; WY Albany; WY Campbell; WY Carbon; WY Converse; WY Fremont
  WY Johnson; WY Laramie; WY Natrona, WY Platte.

Colorado:

157 Denver, CO:
 SMSA Counties:
  2080 Denver-Boulder, CO_____13.8
  CO Adams; CO Arapahoe; C0 Boulder. CO Denver; CO Douglas; CO Gilpin; CO
  Jefferson.
  2670 Fort Collins, CO_____6.9
  CO Larimer.
  3060 Greeley, CO_____13.1
  CO Weld.
 Non-SMSA Counties_____12.8
  CO Cheyenne; CO Clear Creek; CO Elbert CO Grand; CO Kit Carson; CO Logan; CO Morgan;
  CO Park; CO Phillips; :CO Sedgwick; CO Summit; CO Washington; CO Yuma.
158 Colorado Springs-Pueblo, CO:

SMSA Counties:
   1720 Colorado Springs, CO_____10.9
     CO EL Paso; CO Teller.
   6560 Pueblo, CO_____27.5
     CO Pueblo.
   Non-SMSA Counties_____19.0
     CO Alamosa; CO Baca; CO Bent; CO Chaffee; CO Conejos; CO Costilla; CO Crowley; CO
     Custer; CO Fremont; CO Huerfano; CO Kiowa; CO Lake; CO Las Animas; CO Lincoln; CO
     Mineral; CO Otero; CO Prowers; CO Rio Grande; CO Saguache.
159 Grand Junction. CO:
  Non-SMSA Counties_____10.2
     CO Archuleta; CO Delta; CO Dolores; CO Eagle; CO Garfield; CO Gunnison; CO Hinsdale;
     CO La Plata, CO Mesa; CO Moffat; CO Montezuma; CO Montrose; CO Ouray; CO Pitkin; CO
     Rio Blanco; CO Routt; CO San Juan; CO San Miguel; UT Grand; UT San Juan.

New Mexico:

160 Albuquerque, NM:
  SMSA Counties.
   0200 Albuquerque, NM_____38.3
     NM Bernalillo; NM Sandoval.
  Non-SMSA Counties_____45.9
     NM Citron. NM Colfax; NM De Baca; NM Guadalupe; NM San Juan; NM San Miguel;
     NM Santa Fe; NM Socorro; NM Taos; NM Torrance; NM Valencia.

Arizona:

161 Tucson, AZ:
  SMSA Counties:
   8520 Tucson, AZ_____24.1
     AZ Pima.
  Non-SMSA Counties_____27.0
     AZ Cochise; AZ Graham; AZ Greenlee; AZ Santa Cruz.
162 Phoenix, AZ:
  SMSA Counties:
   6200 Phoenix, AZ_____15.8
     AZ Maricopa.
  Non-SMSA Counties_____19.6
     AZ Apache; AZ Coconino; AZ Gila; AZ Mohave; AZ Navajo; AZ Pinal; AZ Yavapai; AZ
     Yuma.

Nevada:

163 Las Vegas, NV:
  SMSA Counties:
   4120 Las Vegas, NV_____13.9
     NV Clark.
  Non-SMSA Counties_____12.6
     NV Esmeralda; NV Lincoln; NV Nye;UT Beaver; UT Garfield; UT Iron;
     UT Kane; UT Washington.
164 Reno, NV:
  SMSA Counties:
   6720 Reno, NV_____8.2
     NV Washoe.
  Non-SMSA Counties_____9.2
     NV Churchill; NV Douglas; NV Elko;NV Eureka; NV Humboldt; NV Lander; NV Lyon;
     NV Mineral; NV Pershing; NV Storey; NV White Pine; NV Carson City

Utah:

165 Salt Lake City, Ogden, UT:
  SMSA Counties
    6520 Provo-Orem, UT _____ 2.4
      UT Utah.
    7160 Salt Lake City-Ogden, UT _____ 6.0
      UT Davis; UT Salt Lake; UT Toole; UT Weber.
  Non-SMSA Counties _____ 5.1
    ID Bear Lake; ID Franklin; ID Oneida; UT Box Elder; UT Cache; UT Carbon; UT
    Daggett; UT Duchesne; UT Emery; UT Juab; UT Millard; UT Morgan; UT Piute; UT
    Rich; UT Sanpete; UT Sevier; UT Summit; UT Uintah -UT Wasatch; UT Wayne; WY
    Lincoln; WY Sublette; WY Sweetwater; WY Uinta.

Idaho:

166 Pocatello-Idaho Falls, ID:
  Non-SMSA Counties _____ 4.0
    ID Bannock; ID Bingham; ID Baline; ID Bonneville; ID Butte; ID Camas; ID Caribou; ID Cassia;
    ID Clark; ID Custer; ID Fremont; ID Gooding; ID Jefferson; ID Jerome; ID Lemini; ID Lincoln; ID
    Madison; ID Minidoka; ID Power; ID Teton; ID Twin Falls; WY Teton.
167 Boise City, ID:
  SMSA Counties:
    1080 Boise City. ID _____ 2.3
    ID Ada.
  Non-SMSA Counties _____ 4.4
    ID Adams; ID Boise; ID Canyon; ID Elmore; ID Gem; ID Owyhee; ID Payette; ID
    Valley; ID Washington; OR Harney; OR Malheur.

Washington:

168 Spokane, WA:
  SMSA Counties:
    7840 Spokane, WA _____ 2.8
    WA Spokane.
  Non-SMSA Counties _____ 3.0
    ID Benewah; ID Bonner; ID Boundary; ID Clearwater; ID Idaho; ID Kootena; ID Latah;
    ID Lewis; ID Nez Perce; ID Shoshone; WA Adams; WA Asotin; WA Columbia; WA
    Ferry; WA Garfield; WA Lincoln; WA Pend Orelle; WA Stevens; WA Whitman.
169 Richland, WA:
  SMSA Counties:
    6740 Richland-Kennewick, WA _____ 5.4
    WA Benton; WA Franklln.
  Non-SMSA Counties _____ 3.8
    OR Baker; OR Gilliam; OR Grant; OR Morrow; OR Umatilla; OR Union;
    OR Wallowa; OR Wheeler; WA Walla Walla.
170 Yakima, WA:
  SMSA Counties:
    9260 Yakima, WA _____ 9.7
    WA Yakima.
  Non-SMSA Counties _____ 7.2
    WA Chelan; WA Douglas; WA Grant; WA Kittitas; WA Okanogan.
171 Seattle, WA:
  SMSA Counties:
    7600 Seattle-Everett, WA _____ 7.2
    WA King; WA Snohomish.
    8200 Tacoma, WA _____ 6.2
    WA Pierce.
  Non-SMSA Counties _____ 6.1

WA Clallarn; WA Grays Harbor; WA Island; WA Jefferson; WA Kitsap;
WA Lewis; WA Mason; WA Pacific; WA San Juan; WA Skaqil; WA
Thurston; WA Whatcom.

Oregon:
  172 Portland, OR:
    SMSA Counties:
     6440 Portland, OR-WA _____ 4.5
      OR Clackamas; OR Muitnomah; OR Washinton; WA Clark.
     7080 Salem OR _____ 2.9
      OR Marion; OR Polk.
    Non-SMSA Counties:
     OR Benton; OR Clatsop; OR Columbia; OR Crook; OR Deschutes; OR Hood River;
     OR Jefferson; OR Lincoln; OR Linn; OR Sherman; OR Tillammok; OR Wasco;
     OR Yamhill; WA Cowlitz; WA Klickitat; WA Skamania; WA Wahkiakum.
  173 Eugene, OR:
    SMSA Counties:
     2400 Eugene-Springfield, OR _____ 2.4
      OR Lane.
    Non-SMSA Counties _____ 2.4
     OR Coos; OR Curry; OR Douglas; OR Jackson; OR Josephine; OR Klamath;
    OR Lake

California:

  174 Redding, CA:
    Non-SMSA Counties_____ 6.8
     CA Lassen; CA Modoc; CA Plumas; CA; Shasta; CA Siskiyou; CA Tehama.
  175 Eureka, CA:
    Non-SMSA Counties_____ 6.6
     CA Del Norte; CA Humboldt; CA Trinity.
  176 San Francisco-Oakland-San Jose, CA:
    SMSA Counties:
     7120 Salinas-Seaside-Monterey, CA_____ 28.9
      CA Monterey.
     7360 San Francisco-Oakland, CA_____ 25.6
      CA Alameda; CA Contra Costa; CA Marin; San Francisco; CA San Mateo.
     7400 San Jose, CA_____ 19.6
      CA Santa Clara.
     7485 Santa Cruz, CA_____ 14.9
      CA Santa Cruz.
     7500 Santa Rosa, CA_____ 9.1
      CA Sonoma.-
     8720 Vallejo-Fairfield-Napa, CA_____ 17.1
      CA Napa; CA Solano.
    Non-SMSA Counties_____ 23.2
     CA Lake; CA Mendocino; CA San Benito.
  177 Sacramento, CA:
    SMSA Counties:
     6920 Sacramento, CA_____ 16.1
      CA Placer; CA Sacramento; CA Yolo.
    Non-SMSA Counties_____ 14.3
     CA Butte; CA Colusa; CA El Dorado; CA Glenn; CA Nevada; CA Sierra;
     CA Sutter; CA Yuba.
  178 Stockton-Modesto, CA:
    SMSA Counties:-
     5170 Modesto, CA_____ 12.3
      CA Stanislaus

8120 Stockton, CA_____24.3
    CA San Joaquin.
Non-SMSA Counties_____19.8
    CA Alpine; CA Amador; CA Calaveras; CA Mariposa; CA Merced, CA Tuolumne.
179 Fresno-Bakersfield, CA:
    SMSA Counties:
        0680 Bakersfield, CA_____19.1
        CA Kent
        2840 Fresno, CA_____26.1
        CA Fresno
    Non-SMSA Counties_____23.6
        CA Kings; CA Madera CA Tulare.
180 Los Angeles, CA:
    SMSA Counties.
        0360 Anaheim-Santa Ana-Garden Grove, CA_____11.9
        CA Orange.
        4480 Los Angeles-Long Beach, CA_____28.3
        CA Los Angeles
        6000 Oxnard-Simi Valley-Ventura, CA_____21.5
        CA Ventura
        6780 Riverside-San Bernardino-Ontario, CA_____19.0
        CA Riverside; CA San Bernadino.
        7480 Santa Barbara-Santa Maria-Lompoc, CA_____19.7
        CA Santa Barbara.
    Non-SMSA Counties_____24.6
        CA Inyo; CA Mono; CA San Luis - Obispo.
181 San Diego, CA:
    SMSA Counties
        7320 San Diego, CA_____16.9
        CA San Diego.
    Non-SMSA Counties_____16.2
        CA Imperial

Alaska:

182 Anchorage, AK:
    SMSA Counties:
        0380 Anchorage, AK_____8.7
        AK Anchorage Division.
    Non-SMSA Counties_____15.1
        AK Aleutian Islands Division; AK Angoon Division; AK Barrow-North Slope Division; AK
        Bethel Division; AK Bristol Bay Borough; AK Bristol Bay Division; AK Cordova
        McCarthy Division; AK Fairbanks Division; AK Haines Division; AK Juneau Division; AK
        Kenai-Cook Inlet Division; AK Ketchikan Division; AK Kobuk Division; AK Kodiak
        Division; AK Kwskokwim Division; AK Matansuska-Susitna Division; AK Nome Division;
        AK Outer Ketchikan Division; AK Prince of Wales Division; AK Seward Division; AK
        Sitka Division; AK Skagaway-Yakutat Division; AK Southeast Fairbanks Division; AK
        Upper Yukon Division; AK Valdez-Citina-Whittier Division; AK Wade Hampton Division;
        AK Wrangell-Petersburg Division; AK Yukon-Koyukuk Division.

Hawaii:

183 Honolulu, HI:
    SMSA Counties:
        3320 Honolulu. HI_____69.1
        HI Honolulu.
    Non-SMSA Counties_____70.4
        HI Hawaii; HI Kauai; HI Maui; HI Kalowao.

---

www.ingramcontent.com/pod-product-compliance
Lightning Source LLC
Chambersburg PA
CBHW081508170526
45166CB00008B/2585